THE WHOLE LANGUAGE COMPANION

THE WHOLE LANGUAGE COMPANION

David Clark Yeager

Scott, Foresman and Company
Glenview, Illinois London

Good Year Books
are available for preschool through grade 12 and for every basic curriculum subject plus many enrichment areas. For more Good Year Books, contact your local bookseller or educational dealer. For a complete catalog with information about other Good Year Books, please write:

Good Year Books
Scott, Foresman and Company
1900 East Lake Avenue
Glenview, Illinois 60025

Printed in the United States of America.

1 2 3 4 5 6 MAL 95 94 93 92 91 90

ISBN 0-673-46188-2

Contents

5 THE SKILLS COMPANION 125

Introduction

The Whole Language Companion is designed to appeal to teachers, students, and everyone who appreciates the strength of the written and spoken word.

The Whole Language Companion is a desktop resource for teachers who want to implement whole language concepts and strategies in their classrooms. Not only does it provide instructors with the background information and rationale for using the whole language approach, but it also includes concise planning and teaching strategies for tailoring the approach to specific classroom populations. Perhaps most importantly, ***The Whole Language Companion*** presents effective student activities that teachers can reproduce and use with little advance preparation.

Moreover, the reproducible activities are not restricted to a particular grade level, are not tied to a designated basal series or literature title, and do not prescribe specific classroom management techniques. Rather, the activities and associated suggestions in ***The Whole Language Companion*** are intended to complement every instructor's strengths and teaching methods.

Many teachers have class-tested the strategies and activities and have found them effective in developing new ideas and instructional approaches.

Five Major Sections

The Whole Language Companion is divided into five major sections. The following paragraphs describe each of these sections.

1: A WHOLE LANGUAGE PRIMER

The first section guides the teacher through the preliminary steps to implementing the whole language approach. It address frequently asked questions such as those dealing with planning, individual and group instruction, and accountability. In addition, it offers clarifying questions and suggestions for creating an appropriate physical environment.

2: THE READING COMPANION

The second section introduces some effective approaches to bringing literature into the classroom. The section focuses on activities related to classroom stories, free-choice reading, interactive reading, and reading associations. In addition to the skill-developing activities for students, this section also offers instructional strategies related to expanding student reading skills.

3: THE WRITING COMPANION
While adhering to the same structure established in the preceding section, the activities in this section focus on the following topics: pre-writing, writer's workbench, writing process, and production and publication. The activities present an opportunity for the teacher to manage a dynamic writing environment in which students can work effectively at various levels of proficiency. In addition, each topic includes instructional strategies that the teacher may use either with small groups or with the entire class.

4: THE SPOKEN LANGUAGE COMPANION
In order to provide teachers with instructional approaches often lacking in traditional language arts programs, this section offers instructional strategies for developing thinking skills and speaking skills as well as methods for creating an optimum environment for student sharing. The activities are specifically designed to assist students in organizing their thoughts and presenting information in a coherent fashion.

5: THE SKILLS COMPANION
This section addresses questions that have been raised about how a whole language approach can integrate the basic skills necessary for proper speech and communication. Not only does it present activities and strategies for student mastery of parts of speech and correct usage, but it also focuses on conferencing techniques that facilitate the integration of basic language arts skills into the whole language environment.

As different as they are, all five sections feature two common elements. Within each section are fascinating "Planning Tips" to add an extra dimension of interest to the topics, and at the end of each section is a "Personal Planning Guide" to spur independent development of whole language strategies.

1 A WHOLE LANGUAGE PRIMER

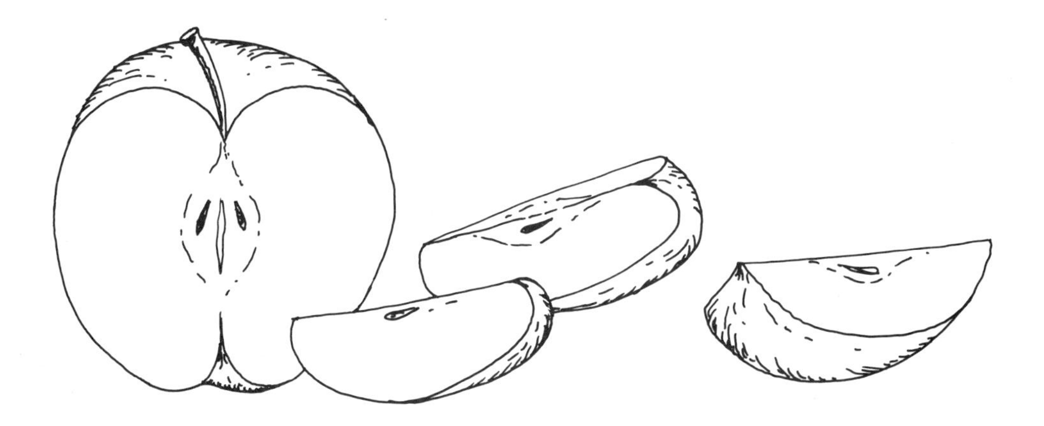

The whole language philosophy of instruction promises to change not only the way instructors teach reading and writing, but also the way they view themselves. Teachers who use the whole language philosophy impart knowledge, of course, but they also actively model the reading and writing process. As a result, their classrooms are marked by the vitality of communication and interaction with language. Whole language teaching fosters an appreciation for language while carefully developing student command of language.

Because whole language is a philosophy of instruction, it is impossible to delineate all the attributes and activities to be found in a whole language classroom. No two such classrooms look precisely the same. Nonetheless, it is possible to point to several traits common to all whole language classrooms:

Students interact to a greater degree with literature.

Students feel an increased ownership of their learning, and they demonstrate a higher level of engagement with that learning.

Teachers relate to each student as both reader and writer, thereby avoiding the "purple flood" of worksheets and skill instruction books.

Teachers demonstrate their appreciation for literature and for writing in general by continually modeling positive approaches to both.

Integrating the whole language philosophy into the regular classroom can seem overwhelming at first—and perhaps even impossible. Fortunately, there are several steps teachers can take to make everyone—students, parents, the administration, and themselves—more comfortable with the direction of the education program and with the inescapable period of transition.

Before attempting to launch a whole language program, a teacher should:

Get to know his or her students. What are their strengths, their fears, their aspirations? Who are their friends? What do they like to do before and after school? What unique experiences do they bring to their learning? The teacher should then use the clues such information provides to select materials that each child may find engaging and to design possible approaches to working with each student.

Become familiar with literature that is appropriate to his or her students. The teacher should develop a partnership with the school librarian (and perhaps even with librarians at local public libraries) and subscribe to periodicals that review new literature for children.

Find out what networks for whole language teachers already exist within the academic community. In doing so, the teacher will likely learn about any conferences at nearby universities that will help facilitate implementation of the whole language approach.

Search for connections within the existing curriculum. What major concepts figure in the teaching of social studies, science, mathematics, and the expressive arts? How can the whole language instructor integrate such concepts into the study of literature? What special projects or events can he or she design around these concepts?

Develop a firm understanding of the school's language arts curriculum. What are students at each grade level expected to know? What reading and writing skills are appropriate for the children with whom the teacher will be working? Although a whole language classroom does not follow a prescribed course of study, the teacher still is responsible for working with children at the level most appropriate to their ability.

Begin to view him- or herself as a reader and a writer. The whole language teacher should keep a journal, recording thoughts, passages, poems, and other original writing. He or she should clip articles and fascinating facts from newspapers, magazines, and other sources. In order that students begin to see their teacher as a reader and a writer, the instructor should begin sharing with the class books read and passages written. Modeling is one of the most effective strategies available to any teacher.

Remember that a whole language program involves structure. In fact, an effective whole language classroom may be more structured than those employing other instructional techniques. In a whole language classroom, students make choices and become partners in their own education; but these choices and this partnership do not lessen the instructor's obligation to provide effective teaching and modeling.

What Does a Whole Language Classroom Look Like?

While no particular physical structure can guarantee a successful program, some arrangements lend themselves better than others to the whole language approach. Most whole language classrooms have several distinct areas:

The Conference Center
It is here that students, teachers, and volunteers come together to discuss books, articles, and one another's writing. Although the conference center generally is not a quiet area, the noise level here should be under control.

The Library
This is where copies of novels, magazines, nonfiction reference books, and published works by members of the class are stored. The noise level in this area is quite low.

The Publishing Center
When students complete a work, they take it to the publishing center, which is well stocked with a variety of papers, pens, markers and other art supplies, and binding materials. The noise level in the publishing center can be high.

Author's Corner
Readings to the class by students, guests, and the teacher all take place here. This area–in which the noise level is focused–should be provided with comfortable seating for all students.

The Reading Center
Often located next to the classroom library, the reading center is where students can relax (e.g., sitting on pillows or in comfortable chairs) while reading to themselves.

The Writing Center
Here students should be able to write without having their concentration disturbed by distractions from other parts of the classroom. The writing center may be nothing more than the students' regular desks or assigned homeroom seats. The noise level will vary here, depending on classroom management styles and expectations.

Planning Tip: You Be the Architect

Before launching the whole language approach in your classroom, take a few moments to examine the physical layout. Where are the traffic patterns? Can you isolate a certain space from the noisier areas of the room or building? Where will your desk be in relation to the various centers you want to establish in your classroom?

Use the grid on the following page to develop a floor plan that will make the best use of your classroom area for a whole language program.

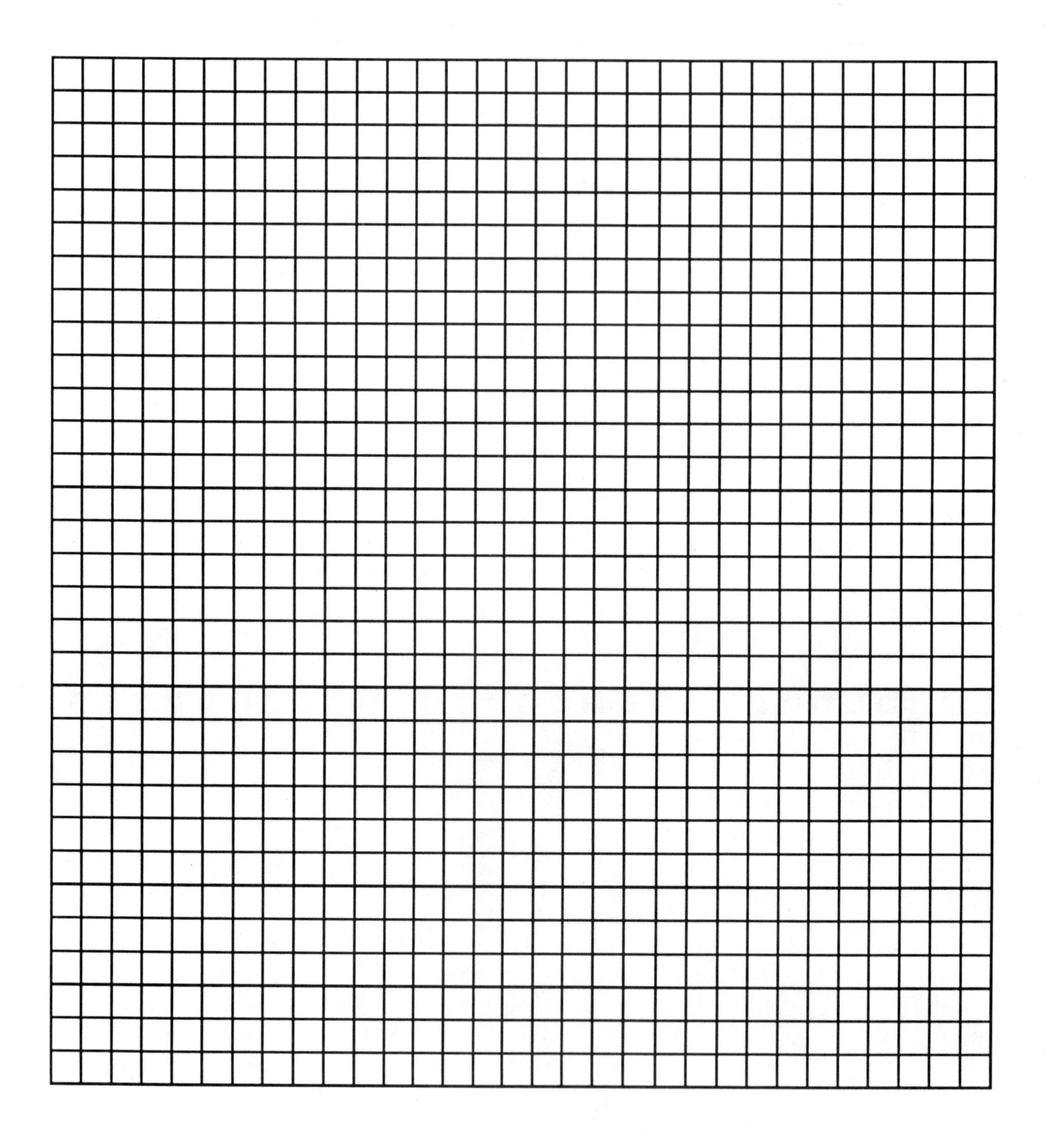

What About Time?

Implementing the whole language approach need not be more time-consuming than administering other forms of instruction. The daily routine in a traditional classroom, for example, is punctuated with periods set aside for basal reading instruction, spelling lessons, handwriting practice, creative writing, and group discussion. In the whole language classroom, all of these topics are integrated with a solid language block. Students develop an expertise in spelling, handwriting, listening, and discussion as they expand their proficiencies in reading and writing. And they do all of this while operating within a routine that the teacher has carefully constructed, orchestrated, and implemented.

As with any model of instruction, the whole language approach has techniques that can enhance and promote effective time management. Naturally, the precise amount of time that any given teacher allocates to the whole language approach will vary according to grade level, student academic and social maturity, the time of day and season of year, as well as the constraints imposed by the overall school schedule. Within that broad framework, however, the teacher who wants a truly effective whole language program should establish a large, undisturbed block of time rather than try to fit the program into several brief and scattered time periods.

Planning Tip: The Right Time

The worksheet on the following page may assist you in structuring your classroom schedule for implementing the whole language approach.

First, block out all periods set aside for lunch, music, art, and physical education as well as any other regularly scheduled events that pull students out of your classroom.

Second, shade in those periods when you know your classroom will be subjected to disturbances from other classes or events taking place in the school.

Third, look at the blank boxes that remain. Do you see a block of at least one hour that you can use for whole language instruction every day? It is preferable (but not necessary) that the block occur at the same time each day.

Where Are Skills Taught?

In the whole language classroom, reading is never taught as a skill from 9:15 to morning recess. In the context of whole language, reading is a valued tool—as indispensable as a pencil—for use all day long. It is through his or her interaction with text that the student gathers, evaluates, critiques, and then shares information. Reading is the primary learning vehicle for both students and teachers.

It is through his or her own reading that the teacher demonstrates the value of reading as a learning tool. The teacher can demonstrate to students—by interacting with text, modeling successful reading techniques, and posing thoughtful questions that reflect the content and intent of the material—how to evaluate critically the words on the page.

This approach doesn't negate the teaching of specific reading skills. Skills are as important in a whole language approach as they are in reading programs based on basal texts. In a basal program, however, the skills instruction is typically packaged in a unit that consists of a reading passage, five to seven comprehension questions, and workbook lessons that often have little relevance to the reading passage.

In the whole language approach, on the other hand, the teacher develops his or her own skills list before the start of the school year. Several good sources exist that provide guidance in this effort. Recent studies on literacy, district curriculum guides, scope and sequence charts from grade-level reading, language books, and, of course, discussions with teachers who have worked with the students previously—all of these sources may be tapped when compiling a list of skills appropriate to the class.

The teacher should act as a diagnostician, selecting those reading skills that need to be taught based on actual experience with the students. Many teachers often stop at this point and remark, "But I have 27 students. If I try to individualize a reading program for each one, I'll never cover the curriculum." Fortunately, helping children become capable readers isn't as complex as might first appear. Several students will likely need help with the same skill, and they can work as a group to master a specific skill. In the whole language approach, students function as a community of learners, sharing their growing expertise and reinforcing what has been taught. It is not only unnecessary but also inefficient to attempt to teach skills to each child in isolation.

How Are Conferences Handled?

For much of the time during the language block, the teacher is moving about the classroom, making suggestions and conferencing with students on an individual and group basis. By asking questions and checking for comprehension, the teacher is assessing the abilities and achievements of each student, thereby formulating a plan for continued success.

Among the many conference models that a teacher might employ, here are three worth considering.

The Individual Reading/Writing Conference

The teacher who uses this model meets with a single student to discuss the child's reading and writing progress and goals. During the conference, teacher and student may discuss a particular book the child is reading or co-evaluate a piece of writing. In the former instance, the teacher assesses the student's response to the book. In the latter case, teacher and student work together to examine the writer's strengths and highlight areas where improvements might be made. Many teachers conclude an individual conference by posing a specific challenge to the student and imposing a definite deadline for the student to meet that challenge.

The Small-Group Conference

In a small-group conference, the teacher and a few students may discuss a particular book or compare several books by the same author or belonging to the same genre. The teacher brings to the conference specific questions to be answered as well as a firm sense of what he or she needs to teach in the small-group setting. At the same time, the students act as partners with both the teacher and one another. They help direct the discussion, each adding to the content and sharing what they have learned. In addition to providing an exciting context in which to learn, the small-group conference is effective in promoting a sense of community within the classroom. Students who become actively involved in sharing what they know tend to be much more accepting of individual differences within the classroom.

The Casual Conference

Here the teacher moves quietly around the classroom, checking for commitment to task as well as proper focus on the work in progress. The teacher may spend two minutes with one child and ten minutes with another. Effective for checking the progress of those students who tend to wander from task and may need a nudge in the right direction, the casual conference is useful in answering the question, "What are they accomplishing?"

Who Is Accountable?

Accountability is always a major concern for teachers, administrators, and parents. It also plays an essential part in the whole language approach. Students who are truly engaged in reading, writing, and sharing what they have learned are active participants in their own education. They tend to hold themselves accountable and responsible for the gains they make, and over the course of the school year they excel at developing the skills and attitudes that will serve them well throughout their academic careers.

This active role for students in no way diminishes the teacher's part in the accountability process. The teacher remains the leader, the one who has a firm understanding of the curriculum, establishes the expectations for success, and sets the appropriate pacing for each student. The whole language classroom is highly structured. Students are expected to read, write, edit, publish, and share. They are expected to build on their strengths, remove obstacles to learning, and take on new challenges. But it is the teacher who keeps everyone on target and who redirects students displaying any sign of falling short of expectations.

Every student has a folder that he or she uses as a "file cabinet" of ideas, responses, articles, and stories. Students draw from this file the material to take through the rough stage, revision editing, and publication processes. It is through conferences that the teacher and student use the contents of this folder to create definite plans for an academic program. It is from this folder that the student takes written material home to share with parents.

Setting the Stage

Before a teacher can move toward implementing the whole language program, several elements must be in place:

Students must be comfortable with a silent reading period. This is not a time for some students to wander around or for the teacher to do attendance. Silent reading is, instead, a time for *everyone* to read. Every member of the class must value the silent reading period.

Each student must have developed the ability to select appropriate literature for him- or herself (or be able to turn to knowledgeable sources for assistance). The responsibility for selecting titles and topics cannot rest entirely with the teacher. If the teacher takes on this burden, students may not take ownership of their learning, thereby losing one of the primary benefits of the whole language approach.

Students must have access to the physical space required to display the products of their learning. The teacher should turn over wall and bulletin board space to student artwork, writing, and research.

Open communication must be established among all participants in the process: teachers, students, parents, and school support staff. If a whole language approach is to prove successful, everyone must be kept up to date on the progress being made.

Planning Tip: Prerequisites

Use the following checklist to help assess what has been accomplished and what remains to be done *before* you implement a whole language approach in your classroom.

____ Are students making good use of the silent reading period? The length of this period will increase with student age and grade level.

____ Are you familiar with all aspects of the curriculum that you are expected to teach?

____ Have you talked about the program with the school librarian (and perhaps with librarians at local public libraries as well as with booksellers)? Have you formed alliances to assist you in developing genre and author lists?

____ Have you discussed the whole language approach with those administrators who are required to evaluate classroom performance?

____ Have parent and community volunteers been trained in conferencing techniques?

____ Have you formulated an architectural plan for the classroom?

____ Have you provided ample space for the display of student work?

____ Are materials available for students to use when they write or create book responses?

____ Have you sent information to parents explaining the whole language program and suggesting how they can establish routines for reading at home?

____ Other: ______________________________________

____

2 THE READING COMPANION

In whole language classrooms—as in traditional ones—students learn the value of reading as a communication tool and spend a large portion of the day reading. They read a variety of material: fiction, nonfiction, news articles, and one another's writing.

One hallmark of the whole language classroom is the way this firm reading foundation encourages children to teach and learn from one another. The teacher acts as a guide and coach for the children. Rather than dividing the class into groups based on standardized reading scores and assigning each group a basal text, the teacher may offer students the choice of several titles within a specific genre or by a single author. The final choice of which book to read may lie with the student. Each child then reads at a rate which is comfortable for him or her. Finally, the student brings what he or she has learned through reading to the group and to individual conferences with the teacher.

The following activities have been designed to promote this sense of community interaction. They allow students and teachers to act as learners, to share their unique strengths, and to create a stage for self-expression. The activities are grouped into four areas:

1. *Classroom Story Activities.*
   The activities in this section are appropriate when a book is being read to the entire class.

2. *Free-Choice Reading Activities.*
   These activities are for the silent reading period. They allow students to track their reading, increase communication between school and home about reading, and promote discussion of books and title recommendations among themselves.

3. *Interactive Reading Activities.*
   The activities in this section are designed to enhance the relationship that children establish with the material they read.

4. *Reading Associations Activities.*
   These activities are intended to be used by several students working in coordination.

Classroom Story Activities

We all know that children love to have stories read to them. Moreover, young children enjoy hearing the same stories again and again so that they can predict (with certainty by the third or fourth reading) what is going to happen next, the characters' reactions, and the outcome of the story. Some even like to simulate taking part in the action of the story. Many of the children who have been read to at home begin school in possession of a variety of reading skills—even though they have never received formal reading instruction.

The classroom story session should be a regular part of the whole language experience. The classroom story session provides an excellent opportunity for students to investigate as a group such concepts as the author's motive and intent, purpose and audience, and use of foreshadowing. It also allows them to begin exploring critical evaluation.

Planning Tip: Coming Attractions

With older students, the importance of selecting the best possible read-along story becomes essential. Fortunately, teachers have access to several outstanding sources of information. In addition to the lists of all-time favorite books published by the American Booksellers Association and local and national right-to-read groups, several periodicals—e.g., *The Horn Book*, *School Library Journal*, *Publishers Weekly*, and the *New York Times Book Review*—publish reviews of new books for children and young adults. Many school libraries and most public libraries subscribe to these periodicals.

Instructional Strategy

Previewing Group Stories

Coming to a Reader Near You

Children arrive at school each year with the same question, "What are we going to learn?" Children love to know—and they should know—what lies ahead for them. If a teacher sits down one day, opens a book, and begins reading, the students have little or no chance to relate the plot and characters to their lives.

This activity provides a simple way to preview for students what they will soon be hearing and reading; it can also provide an effective means for building anticipation. Using materials available at most schools, the teacher can create movie-like posters that advertise coming attractions. In this case, of course, the coming attractions are the books that students will be listening to or reading in the weeks ahead.

To make a large poster showing a scene from the book, use an opaque projector or a photoduplicated transparency of the book's cover or of an internal illustration. Write a short and snappy paragraph about the book somewhere on the poster. If you use an internal illustration for the poster, be sure to note the book's title and author below the drawing, and don't forget to identify the major characters shown.

Student Activity

Character Portrayal

Name________________________ Date________________________

Act I Scene 1

Sometimes when we listen to a story being read to us, the characters and situations seem to come alive in our minds. A very talented author can make us laugh, cry, and grow fearful along with the characters. If someone were to sneak up and tap us on the shoulder in an especially suspenseful moment, we might even jump out of the chair screaming.

Like a good author, a good listener can often capture a moment of high emotion and portray it for others. See if you and a few friends can act out a short scene from the story being read aloud. Use the questions below to help you plan a performance for the rest of the class.

What is the name of the book? ________________________

____

What scene will you be dramatizing? Briefly describe it. ________________________

____

____

____

____

____

Who are the major characters in the scene? List who will be playing each part.

____

____

____

____

____

____

What special scenery or props will you need? ________________________

____

____

____

How long does your performance take? ________________________

____

Instructional Strategy

Predicting Story Events

The Prediction Box

We are always looking to the future, trying to predict what will happen. Children listening to a classroom story do much the same thing. What will the characters do? What emotions do these characters feel now, and how will they feel later? What can they do to solve the book's basic problem?

By asking these questions, you can help children listen for subtle evidence of the changes that are on the way. You can give this process tangible form by starting a prediction box. Every few days, pose one or two questions that require students to predict what will happen in the story you are reading to them. Have them write their predictions on small slips of paper, and then place their predictions in the box (much as you would a ballot).

When you reach the point in the story where the questions you posed are answered, open the box and read the student predictions. The key point here is not to discover which students were able to predict the story correctly. Focus instead on the reasoning that went into the student ideas. You can use both correct and incorrect responses as effective teaching tools.

Instructional Strategy

Class Bibliography

A Web of Titles

One thing always seems to lead to another. When listening to a book read aloud, some student inevitably will say, "That reminds me of . . . ." The student may be reminded of a character in another book, an author's style, or a similar plot pattern. Whatever the similarity, the child has successfully made a connection between two pieces of literature. A child who makes such connections is likely to become a lifelong reader.

One way to reinforce the connecting skill and encourage its further use is to implement the "Web of Titles." Start by placing a poster bearing the title of the book you're reading on the wall. On a table near the wall, place several lengths of string or yarn and a number of index cards like the one below. Each time a student makes a connection between the current title and another book, he or she may complete an index card, attach it to the wall, and connect it to the poster with the string or yarn. By the time you finish reading the current title to the class, you may find a fascinating bibliography on the wall. Encourage students to use the books listed by their classmates for free-choice reading.

Your name: ____

Title of the book you thought of while listening to the classroom story:

____

____

Reasons why you made a connection between these two books:

____

____

____

____

____

____

Personal Planning Guide

What do you want the students to know after they complete these classroom story activities?

____

____

How do you plan to present the information (chalkboard demonstration, A-V presentation, guest speakers . . .)?

____

____

What print materials do you plan to use to support learning?

____

____

Circle the features that you plan to incorporate into these activities:

demonstration application extension discussion evaluation

other: ____

Considering the nature of your students, what is the most appropriate time line for completion of these activities?

____

Complete the time line below by noting the sequence of instruction.

Step I: Introduction of material to students

____

____

____

____

____

Step II: Demonstration and modeling

____

____

____

____

____

Step III: Practice

Step IV: Collaboration/cooperation

Step V: Evaluation

Free-Choice Reading Activities

Students should have as much time as possible during the school day to read, and they should spend a good portion of this time reading materials of their own choosing. It is through their continual interaction with print that students refine the skills they need for success in school.

The value of an extended reading period has been recognized for many years. Some schools refer to this period as USSR (Unstructured Sustained Silent Reading) while others designate it 2SR (Sustained Silent Reading), HIP (High Intensity Reading Period), or LIA (Literacy in Action). Some schools require students to sit silently at their desks while reading; others allow students to bring in pillows on which to read. There are schools that forbid students from reading magazines during the extended reading period while other schools encourage students to read appropriate periodicals.

Despite their differences in naming and structuring the extended reading period, schools that have effectively modeled active reading can point to marked improvement in student reading success. Such success is clearly not the exclusive result of basal reading series and supporting skill lessons. Rather, it evolves from student modeling and practice. When given an opportunity to share what they read with others, students begin developing the ability to evaluate author style, plot structure, and so forth, as well as to become active participants in lively discussions.

The activities in this section are designed to strengthen the bonds between reader and author, student and library, and home and school. The activities may be used in any sequence, simultaneously, or in isolation from one another.

Student Activity

Evaluating Books

Name________________________ Date________________________

Wall of Fame

How many times have you heard a friend complain, "I can't find anything in this library to read!"? How many times have you had the same complaint? Although it may seem at times as though your library has nothing to interest you, it most likely has at least a few real gems just waiting for you to discover them.

The purpose of this activity is to get you and your friends to start sharing your opinions of the books you read. Each time you read a book that you think other students shouldn't miss, take a few minutes to make a contribution to your classroom's "Wall of Fame." Begin by sketching an intriguing book jacket that presents one of the most interesting parts of the book. To the side of the book jacket list a few of the things you liked about the book.

Your comments may inspire someone else in the class to read the book. If that person also likes it, he or she should sign your sketch. As the list of readers grows, add more paper for signatures.

Instructional Strategy

Thematic Reading Collections

The Portable Library

Every teacher has come across students who—despite the best advice, assistance, and coaching—seem unable to find books they think will be of interest. Such students often benefit from being around portable libraries.

Establish several library satellites—places where you can put four or five books on display—around the classroom. These places need not be prominent; look for small spots just below a window or in a little-used corner. Just make certain that each satellite space is accessible to all students.

Once you've located the right places, stock each portable library with titles that have something in common: author, theme, or plot. The idea is to channel reluctant readers toward possible selections by surrounding them with books of a similar nature. Encourage students who read books from the portable libraries to leave brief comments about their selections. Their comments may make it easier for others to select titles of interest.

Be sure to establish some kind of a check-out system so that you know a) where the books have gone, and b) which portable libraries are the most popular. If you find that one or two of the satellites are consistently ignored by students, don't hesitate to introduce entirely new selections of titles to those areas.

Student Activity

A Reading Log

Name________________________ Date________________________

The FCR Log

How many books do you read in a month? a semester? an entire school year? If you were to keep track of the number and subject matter of the books you read, you would learn a great deal about yourself. What you learn might come as quite a surprise!

The form below is designed for you to record *everything* you read. You may want to record just the books you read, both novels and works of nonfiction. But you can also use the form to record newspaper articles, short stories, magazine features—even brochures and other items of information that arrive in the mail. The choice of what to record is entirely yours.

Date	Type of Material	Title

Instructional Strategy

Reading Incentives

The Home-School Connection

Students should never regard reading merely as an academic subject, something to do only when school is in session. Research clearly demonstrates that the most successful young readers are those children who are exposed to books at an early age and who are encouraged to establish a regular reading routine at home.

Many parents would like to initiate a program of home reading, but they are uncomfortable taking the lead role in modeling strong reading habits. Often, however, a gentle nudge from the classroom is all it takes to overcome this parental reluctance.

One effective way to encourage students to read at home is to establish a recognition system based on the total number of hours a child reads outside the classroom. While the possible forms such a system might take are limited only by the imagination, here are a few suggestions:

Ring Around the Classroom: Each ring in a paper chain represents one hour of home reading.

Wall of Readers: Hours are represented by paper bricks that can be stacked to fill a classroom wall.

Reading Railroad: Hours take the form of paper boxcars on an ever-lengthening train.

Supply your students with a recording form to take home. Tell them to mark off one box on the form each time they read for 15 minutes. When a student completely fills all the boxes, he or she should then have a parent or guardian sign the form. It is the completed and signed form that then is transformed into a ring in the chain, a paper brick, or a paper boxcar.

You can modify the form for use with various grade levels. Younger students, for example, may do better with a form that can be completed with just a single hour's worth of reading. Older students, on the other hand, may appreciate the challenge of recording as much as five hours of reading on a single form.

Personal Planning Guide

What do you want the students to know after they complete these free-choice reading activities?

____

____

How do you plan to present the information (chalkboard demonstration, A-V presentation, guest speakers . . .)?

____

____

What print materials do you plan to use to support learning?

____

____

Circle the features that you plan to incorporate into these activities:

demonstration application extension discussion evaluation

other: ____

Considering the nature of your students, what is the most appropriate time line for completion of these activities?

____

Complete the time line below by noting the sequence of instruction.

Step I: Introduction of material to students

____

____

____

____

____

Step II: Demonstration and modeling

____

____

____

____

____

Step III: Practice

Step IV: Collaboration/cooperation

Step V: Evaluation

Interactive Reading Activities

The seemingly simple act of reading is not an end in itself. Rather, it is part of a larger process of sharing ideas with others. Students must be encouraged to take part in that process. Good authors know how to make their readers feel joy, sadness, intrigue, and surprise. By expressing these feelings to others, students capture the essence of the reading experience.

Interactive reading activities are designed to provide students with the opportunity to share reading experiences with their peers and with interested adults at the school. Unlike many commercially prepared book report formats, these activities don't force a structure on the student. Instead, they provide students with points of departure from which to express their feelings and interpretations of a book in a variety of ways.

Planning Tip: Great Literary Events

Consider hosting a "Great Literary Event" at which students can display expressive responses to the books they have read. Often, they will put enormous effort and energy into these responses and derive a valuable learning experience as a result. If possible, hold several of these "Great Literary Events" during the year so that as many students as possible (and perhaps parents, too) have an opportunity to discuss the books they have read. You might even consider inviting local or regional authors to discuss their books, interact with student responses, and provide their responses to other books they've recently read.

Student Activity

Literature on Display

Name________________________ Date________________________

Art's Corner

Reading a good book is a little like stepping right into another world and experiencing all that it has to offer. When you finish reading a first-rate book, you want to talk about it. You want to express what the characters meant to you, how they grew and solved their various problems. You want to tell other people how you felt at different times while you were reading the book.

One way that people share their feelings about a book is by writing book reviews. But here's a way that you can *show* your feelings without writing a single word.

Create a visual display that tells others about the book you read and, especially, how you felt about the book. Be careful, however. Creating an effective visual display can be much more difficult than you think. The questions below should help you plan the right kind of display for the book you read.

What is the title of the book you read?________________________

____

____

What art medium seems best suited to the setting of the book and to the major characters? Circle one of the selections below or add another that is more appropriate to your book.

watercolor macrame paper mache pencil sketch pottery

other:________________________

What materials will you need to create your visual display? List them below and check off each one as you acquire it.

________________ ________________

________________ ________________

________________ ________________

________________ ________________

________________ ________________

How long will it take for you to complete the project?________________________

____

Sketch your design on a separate sheet of paper before you actually begin your visual display.

Student Activity | Sharing Good Reading

Name________________________ Date________________________

A Book Event

Think how much you'd be missing if every time you saw a great movie you couldn't tell anyone what you saw or how much you enjoyed it. The same is true for a great book. When you read a truly excellent book, tell someone about it! Here's one idea for sharing what you read with others.

The purpose of this activity is to help you plan a successful "Book Event." A book event takes place when you turn a portion of the classroom (an area that can be as small as your desk or as large as the entire room) into a scene from the book you just read. Try to make the setting as authentic as possible. If the scene from the book involves trees, draw or construct trees for your set. If the book's characters wear top hats, find or make top hats for the characters in your scene.

Once the book event begins, no one on the set can move! You and the other characters are frozen, just like mannequins in a store window. Try to get as many of your friends as possible to join in the fun. They can help you make your book event a great event.

Before you start building a set, do a little planning for your book event by answering the questions below.

1. What is the title of your book?________________________

____

2. Who is the author of the book?________________________

____

3. In what period of time does the story take place?________________________

____

4. What style of clothing do the book's characters wear?________________________

____

5. Where can you find appropriate costumes?________________________

____

6. What will your set look like (forest, city, desert, outer space, etc.)?________________________

____

7. How many other people will you need?________________________
8. What are the names of people you plan to have in your scene?________________________

____

Use another sheet of paper to sketch your book event set.

Student Activity

Focusing on Genre

Name______________________________ Date______________________

Genre, Genre

If you were reading a mystery, you would no more expect to encounter Martians strolling across the prairie than you would expect to read about cattle rustlers riding down Madison Avenue in a novel about contemporary life. Westerns, science fiction, and contemporary novels are all examples of "genres," or types, of literature. Many readers have a favorite genre while others enjoy books from a variety of genres.

This activity is designed to help you develop a presentation for the class about one genre. The idea is to work with other students who are reading books in the same genre so that your presentation features some aspect of each of the books. If you are reading science fiction, for example, you might want to build a model of a space vehicle. Meanwhile, someone else in your group could act out a passage about life on a distant planet. And a third member of your group could design an instruction manual for an important tool found in his or her science fiction book.

The possibilities are endless. To begin your presentation, you and the other members of your genre group should sit down and plan what each of you would like to do. Group members should then discuss how they might be able to help one another in completing the project.

Start your presentation planning by answering the following questions.

1. Who are the members of your genre group?______________________

2. To what genre do the books in your group belong?______________________

3. What are the titles of the books to be featured in the presentation?______________________

4. How will each title be featured in your group's presentation? Write a short description of what each person in the group plans to do.

   Title 1:____

   ____

   Title 2:____

   ____

   Title 3:____

   ____

   Title 4:____

   ____

   Title 5:____

   ____

   Title 6:____

   ____

7. What materials will your group need to complete this project? Make a list on a separate sheet of paper. Be sure to include everything from paper clips to large sheets of paper.

8. How long will it take your group to finish the project?________________

   ____

Instructional Strategy

Celebrating Books in Character

T(itle) Party

When students complete a book they truly enjoyed, putting it away is a bit like losing a good friend. Through the author's craft, students have come to know the characters intimately—their strengths, weaknesses, fears, and aspirations.

One effective (and enjoyable) way to bid farewell to the characters is to give them a going-away party. Such parties are most effective when several students in the class have recently completed several different titles. After you hold reading conferences with the students—during which they respond to what they have read in an appropriate manner—invite each of them to a special lunch. You might even want to prepare special invitations (see example below).

The only responsibility students have at this lunch is to dress in character and take on the persona of one of the leading figures in the book they just read. Make clear to the students that they should be prepared to answer questions about who they are and what they experienced in their various stories. They should also be prepared to ask questions of the other guests at the lunch.

**You are invited
to a title party
to be held on**

____

**Please come dressed as your favorite
character from the book you just read.**

Personal Planning Guide

What do you want the students to know after they complete these interactive reading activities?

____

____

How do you plan to present the information (chalkboard demonstration, A-V presentation, guest speakers . . .)?

____

____

What print materials do you plan to use to support learning?

____

____

Circle the features that you plan to incorporate into these activities:

demonstration application extension discussion evaluation

other: ______________________________________

Considering the nature of your students, what is the most appropriate time line for completion of these activities?

____

Complete the time line below by noting the sequence of instruction.

Step I: Introduction of material to students

____

____

____

____

____

Step II: Demonstration and modeling

____

____

____

____

____

Step III: Practice

Step IV: Collaboration/cooperation

Step V: Evaluation

Reading Associations Activities

When many of us were in school, we were grouped for reading according to a limited number of criteria: test scores, teacher evaluations from the previous year, and perhaps even the number of books available at each ability level.

In the whole language classroom, students are also grouped, but the selection criteria are entirely different from what we experienced as children. For example, a group today may consist of children brought together because they share a common interest in science fiction (or another genre). Another group may comprise all of the students interested in reading the same book or in sharing the reading of several books by the same author.

These groups in the whole language classroom are often called associations in order to reflect their common interest and purpose. Associations change as student needs, interests, and capabilities change. No association is stagnant.

The precise configuration of each reading association is left to the teacher. Some groups may have as many as 14 or 15 members while others may consist of as few as three or four students. The key to success of each association is not its size but rather the ability of students to meet the group's expectations and to work well within its structure. Group expectations may range from merely completing a reading to sharing critical analysis of a book among peers.

The teacher's role often resembles that of a manager. He or she is responsible for making sure each of the following questions can be answered affirmatively:

Is a convenient meeting place available?
Do the students have a focus for their discussions?

Is there someone within the association (i.e., the teacher or an adult volunteer) who has a sense of the direction in which the group should move and who is able to motivate students to move in that direction?

Are activities available that allow students to express what they have learned in a creative, positive manner?

Is there an open flow of communication between school and home that promotes the association's goals?

If any of these questions cannot be answered with a yes, it is the manager's (teacher's) duty to step in and redirect the association.

Conferencing may be done with an association when the group is very small, but with an association of more than four or five, the teacher should also schedule individual (or at least pairs) conferences. When conferencing with an entire association, the teacher asks general questions about the book—e.g., the plot and characters. Specific questions about a student's ability to read and understand a book should be posed in the individual—not the association—conference. The teacher should schedule personal conferences as needed.

Planning Tip: Members Only

Although associations change frequently and some may last only a short time, you should develop a system for keeping track of each one's members, goals, duration, and accomplishments. You might also want to note which students in the group need extra motivation, guidance, and structure in order to remain effective participants.

One easy system for monitoring reading associations requires nothing more than a set of index cards. The model below illustrates how you can set up your association cards.

Association Focus________________________________

____

Projected Start and End Dates______________________

____

Titles to Be Read_________________________________

____

____

____

Members____

____

____

____

Instructional Strategy

Discussion Skills

Reading Roundtable

The purpose of a reading roundtable is to assist students in discovering that a common ground exists on which they can discuss most literature. You can maximize the benefits of the roundtable discussion by posing several thought-provoking questions. Ideally, you want your role to be that of a participant rather than a discussion leader.

The questions should not be of the sort that evokes a quick yes or no response. Instead, you want questions that elicit responses which in turn invite still more responses and perhaps elaboration of previously suggested observations and opinions. Questions may be philosophical (e.g., "How did the characters strive to improve the quality of life for themselves and those around them?") or practical (e.g., "If the leading character could change just one of his or her actions, what would that action be and how would the change affect the rest of the book?").

Keep a record of the questions you ask. Then, after each session of the reading roundtable, make a few notes about how the students responded to each question. You might also want to jot down ideas for restructuring questions in order to improve student responses.

Consider making up a set of index cards like the one illustrated below to help you maintain your record of roundtable questions.

Question: ______________________________

____

____

Evaluation of Question: ______________________________

____

____

Question: ______________________________

____

____

Evaluation of Question: ______________________________

____

____

Student Activity

Review and Preview of Plot

Name________________________ Date________________________

On the Trail of Something New

Have you ever gone back to a place that you remembered as being big, bright, and colorful only to find it quite small and dark and drab? The strange feeling you experience as you look around the room is caused by the contrast between what you expected to see and what you actually see.

This is a fairly common experience. It even happens when you read. You have certain expectations when you begin a book. Perhaps you've read another book by the same author, or the first characters in the book may remind you of people you know well. Based on your expectations, you think you know how the story will turn out and how the characters will act.

But after reading the book for awhile, you find that what you expected is not taking place. The author whose other books you've read may be writing in a completely different style in this one. Or those characters that seemed so familiar may turn out to be nothing like the people you know so well. Whatever the reason, you find yourself surprised by how different the book is from what you expected.

The purpose of this activity is to help you focus on the unexpected turns that you find when reading. You can share these surprises with the members of your reading association by filling in the blanks below. Then bring this sheet with you to the next association conference.

Book title: ________________________

____

Author: ________________________

Action I thought was going to take place when I began reading: ________

____

____

____

Action that has actually taken place in the story, giving it a completely new direction from what I had expected: ________

____

____

On a separate sheet of paper, write a few short thoughts about what you now predict will happen by the end of the book.

Student Activity

Facts in Fiction

Name________________________ Date________________________

Accuracy, Inc.

In a work of fiction, the author doesn't make up everything from his or her imagination. Good writers include some factual information to add life and color to their work. For example, an author writing about two boys who lived during the American Civil War might have the boys discuss what they want to be when they grow up. But, to have the boys say that they want to be professional baseball players would be a serious mistake. There was no professional baseball in the United States until many years after the Civil War.

Good writers are very careful to create accurate environments for their characters. These environments must truly reflect how things actually were during the time period. You can check an author's accuracy by doing a small amount of research. The purpose of this activity is to help you make such checks for author accuracy.

Book title: ________________________

____

Author: ________________________

Find a passage of the book that includes factual information. This passage might mention a particular location, type of transportation, or a specific activity. Write a brief description of the item discussed in the passage on the lines below.

____

____

____

Find at least two nonfiction books or magazine articles that discuss the same item. On the lines below, write a short summary of the facts about the item as presented in the books or magazines.

____

____

____

____

Do the books or magazine articles support the author's description? On a separate sheet of paper, comment on any important differences you discovered between the fiction author's work and the facts you uncovered during your research.

Instructional Strategy

Reading Logs

Individualized Reading

One of the benefits of the whole language approach is that it allows students to progress at their own rate rather than being forced either to move too quickly or to slow their natural reading pace. A classroom approach that requires every student to be on the same page at the same time often leads to increased anxiety and confusion.

The problem with allowing students to progress on their own is that it makes it difficult for the teacher to assess each student's optimal reading rate. One good way to solve this problem is to provide students with reading logs and with instructions on how to record how much they read within a certain period of time. Make clear, however, that the logs have nothing to do with competition and that students have nothing to gain by entering large numbers of pages on the logs.

Monitor the students so that they fill in their logs each day for at least several weeks. In addition, maintain an annotated record of your own that details how much each student reads and how well he or she functions during conferences. After collecting such data for several weeks, examine the information and then answer the following questions.

Does the student display a good sense of the story?

Does he or she seem to understand the material?

Does the student seem relaxed or anxious reading at this rate?

On days when the student read at a slower rate or read smaller amounts of material, did he or she exhibit greater comprehension?

Use the lines below to devise several other questions that relate specifically to the reading rates and abilities of particular students.

____

____

____

____

____

____

____

Personal Planning Guide

What do you want the students to know after they complete these reading associations activities?

____

____

How do you plan to present the information (chalkboard demonstration, A-V presentation, guest speakers . . .)?

____

____

What print materials do you plan to use to support learning?

____

____

Circle the features that you plan to incorporate into these activities:

demonstration application extension discussion evaluation

other: ____

Considering the nature of your students, what is the most appropriate time line for completion of these activities?

____

Complete the time line below by noting the sequence of instruction.

Step I: Introduction of material to students

____

____

____

____

____

Step II: Demonstration and modeling

____

____

____

____

____

Step III: Practice

Step IV: Collaboration/cooperation

Step V: Evaluation

3 THE WRITING COMPANION

At one time, most student writing in elementary and middle-school classrooms was limited to placing short answers on blank lines. Fortunately, those days have given way to a greater focus on writing. Writing is now seen as a lifelong skill that can be developed only through careful guided practice.

All too often, however, this practice involves worksheets that give students not only the specific topic about which to write but also (in many cases) the first and last sentences. Instead of filling in short answers on blank lines as students did years ago, they merely insert sentences to complete the prescribed activity.

In a whole language classroom, students are allowed and encouraged to choose their writing topic, style, as well as the overall length and format of the piece. This increase in student self-direction does not diminish the teacher's role in developing strong writing skills. The teacher can now concentrate on fostering children's writing through appropriate praise, critical questioning, and sound instruction in the conventions of writing. How all of this can be accomplished in a whole language classroom is the focus of the following section.

Planning Tip: Write Away

As children discover how effective writing can be as a communication technique, many of them become intrigued with the idea of preparing their own nonfiction works. You can enhance their excitement by showing them how to correspond with an expert in the field. Work with the librarian (either at school or at the local public library) to track down directories of nonprofit organizations and government agencies. Show students how to use these directories to create an "Address Hotline" which they can use to track down people possessing an expertise in their research topics. By writing and/or calling for information, students receive valuable practice in business communication as they refine their research and writing skills.

The Reading-Writing Connection

Students in a whole language classroom spend much of their time interacting with text. They may be participating with the teacher in a shared book experience, reading books of a particular author or specific genre with other members of the class, or reading various printed materials of their choosing (novels, nonfiction, periodicals, reference materials) during the silent reading period. Reading is the keystone of the whole language classroom.

What effect does this interaction with the printed word have on student writing skills? When children read, they develop a sense of character, action, and setting. They become active participants with the literature, learning to identify not only with the characters but also with the emotions those characters experience throughout the story. It is in these ways, then, that students forge a connection between reading and writing.

Children can emulate an author's writing style and use that style in their own original stories. For example, children by the thousands mimic the quick narrative style of Judy Blume when crafting their own personal narratives. Not every child has a brother named Fudge or a turtle named Dribble, but they readily see how inclusion of their own siblings and pets can bring color and feeling to their writing. Historical works such as *Johnny Tremaine* or *My Brother Sam is Dead* often prompt children to begin writing with that special sense of urgency that accompanies all social upheaval.

Few people become excellent writers without developing an appreciation of excellence in what they read. One of the primary goals of the whole language teacher, therefore, must be to develop a class of readers. This doesn't mean that every child reads the same book or reads at the same level. It means that the first step in developing creative and capable writers is developing in students a love and appreciation for the written word.

Planning Tip: The Top Three

What are your students' three favorite books? Conduct a poll to find out. Distribute copies of the ballot (see sample below), and ask the students to think carefully about their choices before writing down any titles. Have a box near your desk where students can deposit their completed ballots.

Once you determine the three favorite books, ask how many students in the class have read these books. Determine how many copies you have in your classroom and how many are available in the school library. Encourage students who have read the books to interest others in reading them.

If you teach math to the same group of students, consider graphing the poll results as an integrated lesson.

THE BEST OF BOOKS

My Three Favorite Books Are

1. ______________________________

2. ______________________________

3. ______________________________

The Teacher as Writer

Many people—including some of the world's greatest writers—fear the prospect of facing a blank page. No one is really sure of what he or she is going to write until beginning the writing process. A formidable task for all, it is a paralyzing one for some. And yet, as teachers we expect our students to jump right in and master the writing skill.

Mastery, we now know, goes well beyond knowledge of the mechanics of writing. Developing strong writers involves plenty of practice, guidance, and modeling. Providing practice time and guidance generally presents few problems. But, for many teachers, modeling good writing for their students is a fearsome task.

It is essential, however, for the teacher using the whole language approach to spend part of the day writing while the children write. The writing area should be quiet so that everyone can think, craft, revise, and reformulate. After the writing period, teacher and students should have time to share what they have written. In some classrooms, a special place (e.g., a chair, large pillow, couch) is designated as the author's spot. It is from this spot that writers—including the teacher—read their most recent works to the class.

The following activities are designed to facilitate the three components of writing instruction: practice, guidance, and modeling.

Pre-Writing Activities

Worksheets that ask children to write about their summer vacation or to describe a cloud often encourage young writers to take shortcuts. They excuse students from the laborious tasks of trying to determine what they want to write and what they are capable of writing. Such worksheets define form and function while completely ignoring purpose and audience.

In the whole language classroom, students take charge of the topic selection process. It is not unusual for a student to begin with three or four topics, eventually discarding all but the one topic of choice. While this process is certainly slower than the spoon-feeding technique of the prepared worksheet, it encourages deliberation and it never takes away ownership from where ownership clearly belongs—with the student.

What can a teacher do when a student protests that he or she simply cannot find an appealing topic? The most effective technique lies in prevention rather than cure. Several of the following activities promote "stockpiling" topics for future use. Students who occasionally find themselves without an interesting topic can turn to their topic collections and likely find at least one that suits them. In those relatively few instances in which a student still cannot find a topic of interest, the teacher must then become a coach, helping the student writer discover a promising topic.

Planning Tip: That's the Idea!

Every student should maintain an idea log. This is the child's data bank of information for use in future writing projects. A data bank is never graded. It merely serves as a repository for newspaper clippings, quotes and jokes, short pieces of original writing—anything and everything that a student could derive inspiration and information from during a writing session.

Student Activity

Developing Topic Ideas

Name______________________________ Date______________________________

The Writer's Doodle Sheet

Nearly every writer struggles with choosing a topic. Few people can sit down and know at once what they want to say. One way to deal with this problem is just to jot down whatever comes into your mind. Then you see where your idea leads. This technique has many names: brainstorming, free-form writing, and oratorical doodling.

Try doing some doodling of your own in the space below. Start by writing a word in the center box. Then, in the circles around the center box, enter as many words as you can think of that relate to your first word. Feel free to add as many circles as you need to your doodle sheet.

For example, suppose you write the word "car" in the center box. In the circles you might enter such words as "engine," "tires," "highway," "family," "vacation,"—any words that you can reasonably connect with car.

The next time you find yourself struggling to find a topic to write about, try this technique.

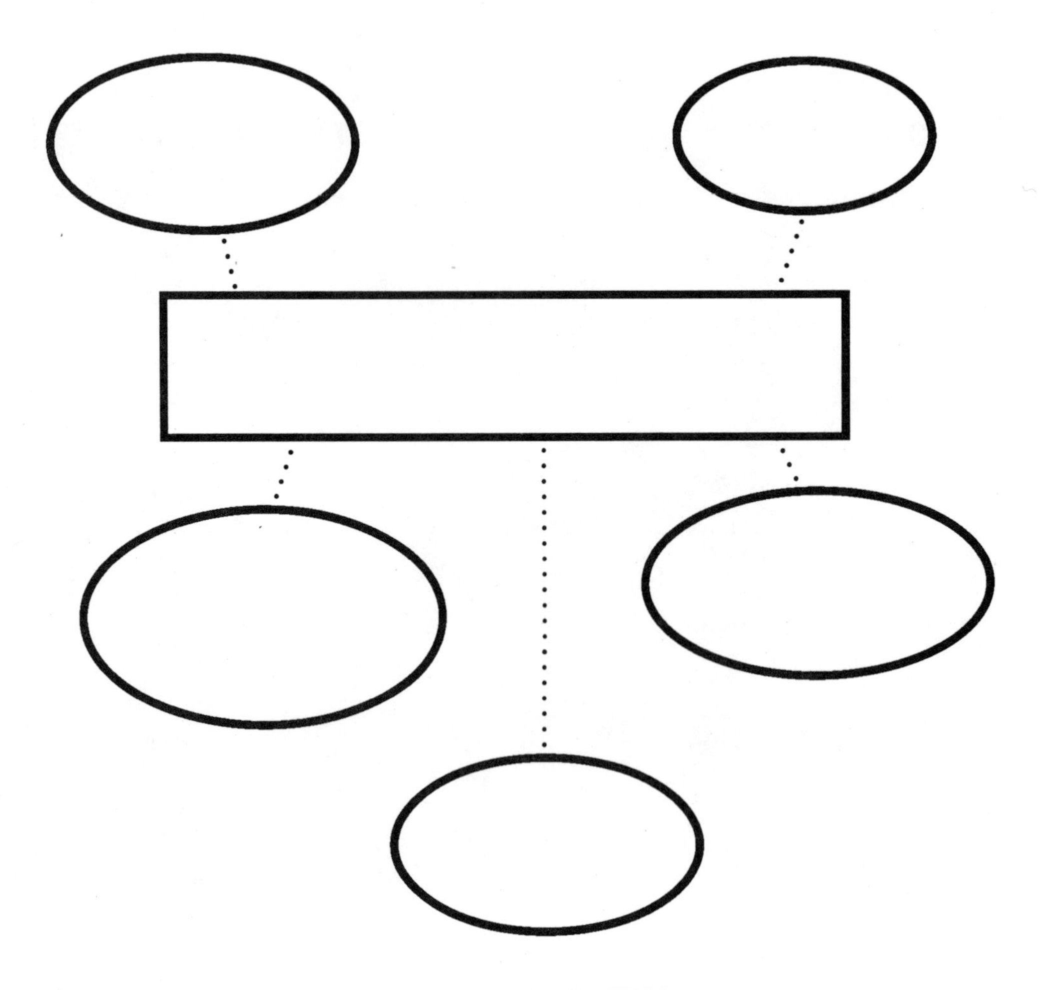

Student Activity

Clarifying Topics

Name____________________ Date____________________

Just the Facts

One of the most challenging jobs at a newspaper is writing a headline that presents an important idea in as few words as possible. A good headline quickly and clearly defines the topic of the story that follows.

Try writing a few headlines for your topic. Use no more than six words in each one. You will probably find that writing headlines for your topic helps you figure what is truly important to your story and what is not. You will also find that the best of your headlines can serve as guides in drafting the story.

Instructional Strategy

Dealing with Writer's Block

Where Do We Go From Here?

Many times, students who have successfully started a piece of writing and progressed nearly through the draft stage hit a block. They then face an agonizing decision: throw the piece away and start over with a new topic or try to muddle through the best they can.

It is at this point that the teacher's role as writing coach becomes invaluable. Students facing a temporary block may benefit from an individual conference to discuss "Where do I go from here?" By exercising a good deal of patience, listening, and guidance—and by utilizing the conference guide below—you may be able to clarify the writer's points of confusion and help overcome a writing block.

Conference Guide

Student ______________________ Date ________ / ______ / ______

Title of work in progress ______________________________

Type of writing ______________________________

What does the student believe is wrong with the piece? ______________________________

____

____

Do you agree with the student's assessment? ______________________________

If you do, what are your immediate ideas for helping the student correct the problem? DO NOT SHARE THESE IDEAS WITH THE STUDENT AT THIS TIME!

____

____

____

____

Ask the student to reread the piece and imagine that it had been written by a friend. What suggestions would the student make to the friend for improving the piece?

____

____

____

____

Compare your ideas for improvement with the student's suggestions. Are any the same? Share similar ones with the student.

If your ideas and the student's suggestions are different, use the student's suggestions to establish the tone and approach for presenting your ideas for correcting the problem.

If the student believes that he or she can move ahead with the piece, what two major ideas for improvement will he or she focus upon?

1. ______________________________

2. ______________________________

How long will the student have to complete the piece, incorporating the agreed-upon changes?

____

Enter a completion date on your master calendar and schedule a follow-up conference for that time.

________ /______ /____

Personal Planning Guide

What do you want the students to know after they complete these pre-writing activities?

____

____

How do you plan to present the information (chalkboard demonstration, A-V presentation, guest speakers . . .)?

____

____

What print materials do you plan to use to support learning?

____

____

Circle the features that you plan to incorporate into these activities:

demonstration application extension discussion evaluation

other: ____

Considering the nature of your students, what is the most appropriate time line for completion of these activities?

____

Complete the time line below by noting the sequence of instruction.

Step I: Introduction of material to students

____

____

____

____

____

Step II: Demonstration and modeling

____

____

____

____

____

Step III: Practice

Step IV: Collaboration/cooperation

Step V: Evaluation

Writer's Workbench Activities

No one would ever expect a mechanic to repair an automobile with nothing more than a Swiss Army knife. Yet teachers often ask students to compose a clear, exciting story with just a pen and paper. For students to become successful authors, they must begin to think like writers. They must carefully craft their words to appeal to a specific audience, create characters with emotions appropriate to the scene, and bring a story to a logical conclusion.

While thinking like an author comes only with considerable experience, the classroom teacher who serves as a writer's coach can foster such success. The following activities provide the structure necessary to coach emerging authors successfully. The activities are not intended to be used with every student in the class. Rather, they are designed for those students who need some extra help in a specific area.

Because they provide guidance rather than direct instruction in developing a specific writing skill, these activities may be used more than once with the same student. In fact, students who do these activities should keep their completed sheets in their writing folder. Then they can refer to the sheets whenever similar needs arise as they work on future articles and stories.

Planning Tip: How Would They Have Said It?

Children often emulate favorite authors, trying to adopt a style and make it their own. This is a phenomenon that teachers should encourage, especially with students who are having difficulty developing mood or tone in their stories. By closely examining the works of favorite authors, students frequently can spot techniques used to bring feeling to a passage.

If you discover such emulation in your classroom, make a note of both the student and the author involved.

Student Activity

Determining Audience

Name______________________ Date______________________

Is Anyone Out There?

Nothing is more disheartening to an author who has spent a great deal of time and effort writing, editing, and polishing a story or an article to find that no one is interested in reading it. Few ten-year-olds, for example, would be interested in reading a story about two elderly women on a trip to Paris—no matter how well the story was written. That's why it's important to determine the audience for your work whenever you begin planning any piece of writing.

If you've already chosen a topic but are worried about whether the subject will find an approving audience, spend a few minutes answering the questions below.

What is the title of your article or story? ______________________

____

Can you name three people who will enjoy reading your piece of writing?

1. ______________________
2. ______________________
3. ______________________

What styles of writing do these three people like to read? Ask each of the people listed above to name three styles they enjoy.

1. __________ __________ __________
2. __________ __________ __________
3. __________ __________ __________

Are there similarities among the styles listed? ______________________

Does your story or article match the kind of writing that your audience likes to read? ______________________

If it doesn't match, what can you do to refocus your writing so that it appeals to your audience? ______________________

____

____

If it is not possible to refocus your story or article, can you change your audience? ______________________

What is the setting of your story or article? ______

______

Will this setting appeal to your audience? ______

What can you do to make the setting more appealing to your audience?

______

______

______

______

Who are the major characters in your story or article? Describe one or two of them. ______

______

______

______

______

______

Will your audience be able to understand and relate to these characters?

______

What can you do to increase the appeal of these characters to your readers?

______

______

______

______

______

Is the pacing of your story or article (meaning how quickly or slowly the action takes place) right for your audience? ______

______

What adjustments can you make when you edit your writing to improve the pacing? ______

______

______

______

______

Student Activity

Extending Plot and Action

Name________________________________ Date____________________

There's More to the Story

Most writers start out thinking that what they are creating is great. About halfway through the story, however, they often begin to lose steam. Then they suddenly face the question: "Where do I go from here?"

This activity is designed to help you review your story and develop different ways to move it toward a successful conclusion.

Story title: ____

Is the story fiction or nonfiction? ____________________________

What is the story's first sentence? Copy the sentence here. __________

____

____

____

Briefly describe the story's action as far as you've written it.

____

____

____

____

____

What are three different ways you could end the story? Use another sheet of paper if necessary to describe three logical endings for the story's action so far.

1. ____

____

____

2. ____

____

____

3. ____

____

____

Which one of these endings makes the most sense? Before choosing one, think about your reason for writing the story, the audience likely to read it, and what you want your readers to gain from reading it.

____

____

____

____

How can you tie your ending to what you've written so far? Be sure to include changes in your characters, setting, and pacing.

____

____

____

____

____

____

On a separate sheet of paper, write a brief outline of the story as it moves toward its conclusion.

Student Activity

Balancing Dialogue

Name______________________ Date______________________

He Said, She Said

In a well-written story, the characters are living breathing people who communicate with one another. Their conversations are called dialogue. Well-written dialogue can make a story exciting, and it keeps the action moving. On the other hand, stories that have too much dialogue can be just as difficult to read as stories that have too little. Good writing requires a balance between dialogue and narrative.

One common mistake that new writers make when crafting dialogue is to end each character's comments with *he said* or *she said*. This activity is designed to help you look at the balance between narrative and dialogue in your writing and to help you avoid constant repetition of he said and she said.

Story title: ______________________

Does your story have dialogue? ______________________

How many characters speak in your story? ______________________

Do some of your characters speak with an accent or use special language that gives the reader an idea of their age, sex, or level of education?

____

____

____

____

Can you improve the dialogue to better show their special characteristics? Make notes to yourself here on ways to improve your characters' dialogue.

____

____

____

____

____

____

Do you have the right balance between narration and dialogue?

____

____

Does your story describe the situations in which conversations take place, how the characters move as they speak to one another, and any distracting things that take place around the characters as they speak? Make notes here about places you can add descriptive information to create a better balance between narration and dialogue.

How many times have you included *he said* and *she said* in your dialogue?

What words could you substitute for *said*?

Can you replace any of the *he said/she said* dialogue endings with more interesting descriptions?

Student Activity

Developing Character

Name_______________________ Date_______________________

Emotion in Motion

We usually can tell when people are upset, happy, outgoing, frightened, or hateful. Their emotions are visible in the way they walk, talk, and interact with the world. The same is true of characters in writing. The way an author describes his or her characters through the action of the story tells us a lot about how the characters see and feel about the world.

This activity is designed to help you recognize whether you are letting readers know what your characters are feeling. It should also help you develop an inventory of words to use for expressing certain emotions.

Story title: _______________________

Use one of the boxes below for each of your major characters. At the top left, write the name of the character. At the top right, enter one word that describes the main emotion the character shows in your story. Then, in the large area of each box, write as many words as you can think of that would show this emotion to a reader. Draw as many boxes as you need for your story.

Instructional Strategy

Eliminating Excess Verbiage

Trimming the Excess

For many young writers, being told to cut a word can feel like an indictment of the entire piece. Helping students make the transition from writer to editor requires patience and positive coaching.

In many cases, a small group conference works better than individual conferences in getting students to eliminate verbiage and begin to write succinctly. Talk to the group in general terms about tightening their writing, and then direct students to find examples of excess in their own articles and stories. Most will be able to find such examples quickly. Encourage these students to revise their work while you help others learn to recognize excess wordage.

The self-editing guidelines that follow are designed to help both groups of students: those who can quickly identify verbosity in their writing and those who see no problem with too many words.

1. Go through your story or article. Lightly underline those words, phrases, and sentences that could be eliminated without changing the tone or meaning of your writing.
2. Reread your story or article, this time leaving out the underlined portions. Does your writing sound better, worse, or about the same? If it sounds worse, go back to step 1 and re-evaluate what you underlined.
3. Read your story or article to a friend. Ask your friend whether he or she agrees that the changes improve your writing. If the answer is yes, rewrite the piece so that it no longer includes the underlined words, phrases, and sentences.

Personal Planning Guide

What do you want the students to know after they complete these writer's workbench activities?

How do you plan to present the information (chalkboard demonstration, A-V presentation, guest speakers . . .)?

What print materials do you plan to use to support learning?

Circle the features that you plan to incorporate into these activities:

demonstration application extension discussion evaluation

other:

Considering the nature of your students, what is the most appropriate time line for completion of these activities?

Complete the time line below by noting the sequence of instruction.

Step I: Introduction of material to students

Step II: Demonstration and modeling

Step III: Practice

Step IV: Collaboration/cooperation

Step V: Evaluation

Writing Process Activities

The writing process frequently involves three stages. During the first (or pre-writing) stage, students begin to think like writers. They develop their own (sometimes humorous) writing habits. They may select a favorite pen or pencil to use when writing. Quite often, they must have their workspace arranged "just so": paper in one place, perhaps a small stuffed animal in another, and a dictionary within easy reach. Or they may assume a unique posture as they think about what they intend to write. Try to allow for such individual preferences whenever possible.

During the second stage, the developing author begins to identify what he or she likes and dislikes about the works of published authors. The student begins to read with a critical eye as well as for pleasure, and he or she starts making notes about scenes and anecdotes that seem worth remembering. The student's idea log fills with a variety of information: news clippings, cartoons, and perhaps even comments from one student to another. The incipient writer may use some of this information in future writing and may never look at the rest of it again. But all of it belongs to the writing process.

During the third stage, the student begins writing. It is at this point—with the student recording thoughts and emotions, crafting a story—that the teacher can serve most effectively as a writer's coach. The coach's main role is to encourage and guide the writer, not to set deadlines and rush the writer through revisions. Students in the third stage of the writing process need to be reminded and reassured that they will not be graded on the basis of the first thing they put down on paper. The teacher/coach should help students recognize the benefits of crafting, reading, evaluating, rewriting, and polishing their writing.

The following activities are designed to assist students as they move through the writing process.

Student Activity

Noting Changes for Future Revision

Name______________________ Date______________________

The Author-Editor's Notebook

Many writers keep a notebook (or several of them) in which they record information, notes about stories on which they are working, descriptions of characters, and ideas for improving a particular scene. When they complete a rough draft and begin revising a story, writers often go back to their notebooks to remind themselves of the possible changes they had thought about earlier.

Some authors even keep a notebook near their beds so that they can record any thoughts or ideas that may come to them during the night. Although you need not take a notebook with you everywhere you go, it's good to get in the habit of writing down all the interesting ideas that pop into your head.

For the next few weeks, maintain a notebook in which you write down fascinating facts, portions of sentences, sketches and doodles—everything you think about that relates to your writing. Review your notebook just before you start to edit your next story or article. The material you've written down may help you improve the final draft.

Here's an example of what a page in your notebook might look like:

January 2, 1990

In "A Train Going West," I think that Seth should have a better sense of humor. On page 2, I have him sitting next to a Union soldier who is telling funny stories about life back home. Seth should show that he appreciates the humor instead of just responding, "Yes sir" or "No sir."

Call the historical society and try to get a copy of an old train schedule. What did the schedule look like? What color paper was it printed on?

Watch for the correct spelling of "Tennessee."

Student Activity

Visual Revision

Name________________________________ Date________________________

Sketch and Scratch

Writing is a *recursive* process. Recursive means that when you write, you constantly go back and reread, rewrite, and re-evaluate what you have already put down on paper. This process helps you make sure that what you have written is what you really want to say.

If you were to spend all of your time rereading and rewriting, however, you'd have a tough time ever reaching the end of the story. This activity is designed to help you make changes without spending too much time in the recursive process.

When rereading your story, sketch the action as if you were telling it in a cartoon strip. You can use the boxes below for your next story. When you finish sketching the action, read the cartoon strip to yourself. Do things take place in the right sequence? Do the characters' actions make sense to you? If you can't answer yes to these questions, go back to your story and scratch out the parts with problems. Make all the necessary changes.

Student Activity

Editing Marks

Name________________________ Date________________________

Proof Positive

Editors and proofreaders use symbols to indicate changes they think should be made. These symbols, called proofreader's marks, can save a great deal of time. Instead of writing out a message that describes the change, the editor or proofreader simply marks the text with a symbol. The author, typesetter, or printer then can make the change.

You can find a list of proofreader's marks in many dictionaries. Several of the most common ones are shown below. Try to get in the habit of using these symbols when you edit your own work or some other author's writing.

Symbol	Meaning	Example
(delete mark)	Delete	the large green monster
^	Insert	the ^ gila monster (large)
≡	Capitalize	bob's new car
/	Lower case	Bob's new Car
(transpose mark)	Transpose	the monster green
(close-up mark)	Close up space	the g reen monster
(sp)	Spell out word	Bob's (2) new cars
#	Insert space	This is Bob'snew car.
⊙	Insert period	This is Bob's new car ⊙
∨" ∨"	Insert quotation marks	"Look at Bob's new car, ∨" said the green monster.

Instructional Strategy

Sharing Written Works

Arthur's Chair

When given an opportunity to read their own work aloud, most children are at first reticent, then cautious, and eventually eager. At the beginning of the year, you likely will have a class of reluctant performers. By spring break, however, you probably will find that most of your authors can't wait to read their writings to their peers.

Providing young writers with the opportunity to read their material to the class is an important part of the whole language approach. Unfortunately, finding the time to allow everyone to read who wishes to do so can be a problem.

A student who reads his or her material should be seated in a place of honor. In some classes, this spot is designated the "Author's Chair." In one classroom, it came to be called "Arthur's Chair" in honor of a student who had a particularly difficult time getting anything down on paper. When he eventually succeeded, however, he earned the respect of his classmates.

Whatever you call the place of honor in your classroom, make certain that it does not become monopolized by a handful of very good writers. Here are some suggestions for managing this area and assuring widespread use of "Arthur's Chair."

Can It: Give each child a tongue depressor, a specially marked pencil, a small flag attached to a short shaft, or some other object that can be marked so that its owner can be readily identified. Place an empty coffee can in a prominent place near "Arthur's Chair." Whenever a student wishes to read, he or she places the object in the can. If several objects are placed in the can, names can be drawn at random to decide the sequence in which students will read. The teacher keeps the objects of those students who have read so that there will be no repeat performances until everyone who wants to read has had a chance.

Readers' Tea: At the beginning of the week, post sign-up sheets—one for each day of the week. A student who wishes to read puts his or her name on one of the sheets. If no space remains on the day a student wants to read, he or she must move to a second or even a third choice. If no space remains on any of the sheets, the student must wait for the next cycle the following week. Students may sign up to read more than once only if space is still available on the sheet for that day. Be sure to hold two or three spaces open on days later in the week for possible spill-overs from previous days.

If possible, have one of the student readers each day bring a small snack to share with his or her classmates. Snacks always follow the reading session.

By Invitation Only: After finding out who wants to read and figuring out an effective order of performances, send an invitation to each student reader. On the invitation, be sure to list the student's name and the title of his or her work. Although you may need to prepare invitations several times a week, make each one as meaningful as possible. To the student author, your invitation will become a memento of his or her achievement in writing.

Personal Planning Guide

What do you want the students to know after they complete these writing process activities?

____

____

How do you plan to present the information (chalkboard demonstration, A-V presentation, guest speakers . . .)?

____

____

What print materials do you plan to use to support learning?

____

____

Circle the features that you plan to incorporate into these activities:

demonstration application extension discussion evaluation

other: ____

Considering the nature of your students, what is the most appropriate time line for completion of these activities?

____

Complete the time line below by noting the sequence of instruction.

Step I: Introduction of material to students

____

____

____

____

____

Step II: Demonstration and modeling

____

____

____

____

____

Step III: Practice

Step IV: Collaboration/cooperation

Step V: Evaluation

Production and Publication Activities

The day has arrived! After hours (or weeks or months) of trial and error, crafting and revising, the student is finally ready to publish a finished manuscript. The significance of this moment rests entirely in what the writer has achieved in personal terms, not in the length or complexity of the work itself. One writer may be ready to publish a novelette while another may have prepared a set of instructions for teaching a new dog old tricks. Both students should be praised for their individual achievements.

Students who are ready to publish something they've written should be encouraged to dedicate the work to someone important in their lives. Including a dedication is a good idea for two reasons: 1) It promotes a feeling of ownership, a feeling on the part of the writer that he or she has produced something of value, and 2) a dedication makes the work a gift to another person, and gifts often elicit greater respect than most classroom assignments receive. Every piece of writing that a student reads to the class should begin with the words, "This is dedicated to . . . ."

Actual physical production of the published work may take several forms. The most common include a paperback edition for the classroom library, a bound edition for parents, and a large illustrated wall story suitable for mounting as a poster. Some of the following activities are designed to reinforce these common publication methods while others introduce new ways to publish a finished piece of writing.

Planning Tip:
© (The Copyright Symbol)

At the beginning of the year, discuss the meaning of the copyright symbol with your students. Point out that the symbol stands for the legal protection that authors enjoy. This protection makes it illegal for anyone else to use an author's work without first getting the author's permission. Encourage your students to place the copyright symbol on their articles and stories, and then issue them copyright certificates.

Student Activity

Moving from Text to Performance

Name______________________ Date______________________

On Stage

Not every author sees his or her creation performed on the stage or movie screen, but those who do experience an enormous feeling of success. Some authors, in fact, work specifically for this goal. They write plays for the stage or screenplays for movies. Authors who write short stories and novels must have their works changed into scripts that actors can perform.

Take one of your stories and rewrite it as a play that you can perform with several of your friends. Keep in mind that nothing can happen on stage unless you include it in the script. That means you must indicate what all the people on stage are supposed to *do* as well as what they are supposed to *say*.

For example, suppose that your character named Christopher is supposed to cross the stage and ask a police officer for directions. Your script might look like this:

CHRISTOPHER (*Crossing from stage left to stand face to face with Officer Mulroney*): Can you give me directions to 777 Florissant Drive?

If several bystanders are supposed to overhear Christopher's question and react in some way (fear, anger, surprise, disbelief), your script must also make their action clear.

Bystanders overhear question and react fearfully, immediately whispering among themselves. Christopher notices their reaction and looks confused.

Now it is time for Officer Mulroney to respond. Your script must spell out how he is to tell the audience what he is thinking. His words and actions must make clear why the bystanders reacted with fear.

OFFICER MULRONEY (*Looking at Christopher as if he were some sort of criminal not to be trusted*): 777 Florissant Drive, eh me boy? Suppose you tell old Officer Mulroney just what a nice boy like you would be doin' in a place like that?

In addition to the specific instructions in the script above, there are several common stage directions that people who write plays use all the time.

The list below presents some of these directions and gives their meanings.

Enter stage right: Actor walks onto stage from his/her right.
Enter stage left: Actor walks onto stage from his/her left.
Exit stage right: Actor departs stage to his/her right.
Exit stage left: Actor departs stage to his/her left.
Aside: Actor addresses the audience rather than another character on stage.
Style of speech: Actor speaks in a certain manner—happily, quickly, slowly, etc.
Act: This is a major division in a play, like a chapter in a book. Most plays are divided into three acts.
Scene: This is a smaller division in a play. Each act may contain several scenes. At the beginning of each scene, the playwright describes what the stage should look like, what props are on stage, where the actors are standing, and how the stage is lighted.

Student Activity

Book Design

Name____________________________ Date____________________

Readers' and Writers' Press

There comes a thrilling time in every writer's life when it's time to say, "The piece is finished!" This is the time to start thinking about how the work will look when it is published. Publication can be the most exciting part of the writing process.

Publication involves many questions that must be answered. Here are a few of them:

What format is best for my material?
Should my work be illustrated?
Can I do the illustrations, or should I ask someone with greater artistic ability to do them for me?
Should the text be typed or hand printed?
What kind of cover would make people interested in my work?

The following questionnaire is intended to help you get ready to publish your work. By taking a few minutes to answer the questions, you'll find yourself developing a mental picture of how you want your published writing to look. If necessary, use an additional sheet of paper to sketch various cover illustrations. Remember, *you* are the publisher. You can design your work any way you like!

What is the title of your work?____________________

Is it fiction or nonfiction?____________________

To whom will you dedicate your work?____________________

Will the text be typed or hand printed?____________________

What kind and color of paper will you use?____________________

Will your work be illustrated?____________________

If it will be illustrated, will the drawings be in color or black and white?

____

____

How many illustrations will be included? ______________________

____

Who is the illustrator? ______________________________

____

When must all the illustrations be completed? ___________________

____

How many pages will your work be (including illustrations)? ________

____

What kind of illustrations do you want on the cover? _____________

____

Student Activity

Book Promotion

Name________________________ Date________________________

The Book Jacket Blurb

After a book is published, the author and publisher don't just sit back and hope people will buy it. Working together, the author and publisher promote the book by telling readers why they would enjoy reading it. Some of this promotion takes the form of advertisements in newspapers and magazines and author interviews on radio and television.

In addition, the book jacket itself is often designed to promote the book. Eye-catching illustrations along with hints about the plot, setting, and characters are intended to encourage readers to pick up the book and start reading. Often these hints are given in the form of "blurbs"—short passages taken directly from the text. Selecting blurbs for the jacket involves reading through the text to find passages that will grab the reader's attention without giving away too much information.

This activity should help you design a book jacket and select blurbs. Use the outline of a jacket on the following page to sketch an illustration on the front cover and to print blurbs on the back cover. Remember to include the book's title, author, and illustrator on the front cover.

In addition to blurbs on the back cover, you can also include favorable comments by people who have read the book. Ask a few of your friends who have read your work if they would be willing to be quoted on your book jacket. If so, write down exactly what they say and print their comments on the back cover. Be sure to include each person's name on the back cover very near his or her comments about your book.

Instructional Strategy

Celebrating Publication

The Book Signing

As an incentive to would-be authors, consider holding a book signing celebration every time one of your students "publishes" a book. A book signing is a great way to reward your more prolific writers as well as a useful motivating tool for those who are struggling to bring their projects to completion.

When a student completes a final edited manuscript, run off several clean copies and bind each one in an attractive cover. Present the bound copies to the author shortly before the book signing. At the event itself, the author signs his or her name on the first page and then distributes the bound copies to the guests. The author's autograph makes each of the books a collector's edition.

Personal Planning Guide

What do you want the students to know after they complete these production and publication activities?

____

____

How do you plan to present the information (chalkboard demonstration, A-V presentation, guest speakers . . .)?

____

____

What print materials do you plan to use to support learning?

____

____

Circle the features that you plan to incorporate into these activities:

demonstration   application   extension   discussion   evaluation

other: ______________________________

Considering the nature of your students, what is the most appropriate time line for completion of these activities?

____

Complete the time line below by noting the sequence of instruction.

Step I: Introduction of material to students

____

____

____

____

____

Step II: Demonstration and modeling

____

____

____

____

____

Step III: Practice

Step IV: Collaboration/cooperation

Step V: Evaluation

4 THE SPOKEN LANGUAGE COMPANION

Research indicates that in the traditional classroom, the teacher spends far more time speaking than do the students. Whether the teacher is instructing the class as a whole or giving directions to small groups of students, the teacher is the focal point of the learning process.

In the whole language classroom, the situation is far different. Everyone shares the roles of teacher and learner, and everyone has the responsibility to present information. Every student is encouraged to articulate his or her views and opinions. At the same time, every student comes to understand that *listening* is an essential component of spoken language skills.

The verbal sharing of ideas is so important in the whole language classroom that often a specific area of the room is designated the "conference center" or "discussion den." It is in this area that students come to share ideas with one another, read passages they think need improvement, and seek advice on editing methods. Although a great deal of trial and error takes place in the conference center, the primary purpose of this area is to facilitate constructive criticism—not chaos. The whole language teacher must establish guidelines for the proper use of the area early in the year and make clear that he or she will show little tolerance for any disruptive behavior.

The success of the conference area often depends on the existence of two conditions. First, the teacher must model strong listening and presentation skills as well as effective questioning strategies. Second, any student who wishes to use the conference area must have something specific to share with his or her peers. Reciprocity is a key element in the success of the conference center. Students give and receive advice. No one goes to the conference center merely to "dish out" criticism or praise.

To enhance the development of these two conditions, the teacher may use several strategies. He or she may read a short piece of writing to the class. Before reading the piece, however, the teacher conducts a discussion of common etiquette and the difference between constructive and negative criticism. This discussion should emphasize the expectation that students will treat each other with respect—if not friendship.

After reading the article or story, the teacher launches a discussion of the praiseworthy aspects of the writing as well as places in which the writing might be improved. The teacher lists the positive and negative points on the board or on an overhead projector so that everyone can see and hear the major ideas being discussed. This is the time for the teacher to emphasize that students ask questions in a positive way, and it is also the time to encourage students to seek alternatives to poorly phrased questions. If some students seem unable to make appropriate comments or suggestions, the teacher might recommend a bit of role-playing. What kind of comments would these students like to receive if they were the author of the article or story?

Another use of spoken language skills in the whole language classroom takes place during the production stage of writing. Many students enjoy transforming their written work into a short play or larger dramatic production. They can utilize the acting skills of their peers, watching and learning as the actors take the written word and give it life on the stage.

Planning Tip: C3B4 Me

One of the most difficult tasks for teachers just starting to use the whole language approach is sharing control with the students while still directing the learning process. One effective way to accomplish this feat is to establish a "See Three Before Me" chart.

Construct the chart by listing various areas of language skill proficiency—e.g., spelling, pronunciation, grammar. Students who have demonstrated their capabilities in one or more of these areas then place their names beneath the appropriate heading(s). Then, when a student has a problem, he or she consults the chart and attempts to get the problem resolved with the help of one or more of the student "experts." If the student with the problem still needs help after consulting three peers, he or she then seeks the teacher's assistance.

The Teacher as Listener

"If I could say things just once—or even twice—and have students understand, I would be happy," a fifth grade teacher complained. "It seems that no matter what I do, there is always at least one student who isn't listening. Sometimes I feel as if I'm talking to the wall."

What this highly qualified and generally successful teacher doesn't realize is that she *is* talking to the wall. It is the wall that she built by modeling poor listening habits.

There is an old proverb: He who listens ill answers ill. Unfortunately, the truth of this adage is apparent in classrooms across the country. Teachers often assume they know what a child is going to ask before the student has a chance of verbalize the question. Worse, some teachers will answer the student's unasked question with inappropriate, incomplete, or poorly thought out information. Students who are repeatedly exposed to such bad listening habits tend to acquire the bad habits themselves.

In an effective classroom, the teacher places as much emphasis on listening skills as he or she does on reading or writing skills. Listening is an essential tool for learning, and good listening habits must be fostered and reinforced whenever possible. The best way to create a classroom of good listeners is to model effective listening traits. Wait for the entire question to be asked before offering an answer. Summarize what you hear, making certain that you and the speaker share the same essential information. Restate the question if necessary. Encourage others to join the exchange of ideas whenever appropriate. Remind students that becoming a capable listener is as important as becoming an effective speaker.

Planning Tip: Hear Ye! Hear Ye!

Without letting students know what you are doing, keep a record of who uses effective communication techniques in the classroom. Maintain the record for an entire week, noting which students are effective communicators and what strategies they use. At the end of the week, share your observations with the class. Make a chart of the most successful techniques, and encourage your students to try them.

Group Participation

In a whole language classroom, students spend much of their time working in groups. Some of the groups are informal (such as students working in pairs in the conference center or several students acting in another's play) while other groups are well defined, brought together by the teacher for a specific purpose. Groups may meet for a direct instruction session—often called mini-lessons, focus meetings, or skill sessions—or in the more informal topic conferences during which students discuss with the teacher how a particular skill might improve their writing.

Group work helps students develop positive social attitudes, reinforces specific language skills, and assists the teacher in providing the best instruction possible. Over the course of the school year, every student will belong to several different groups. A student who needs instruction, reinforcement, or remediation in a specific language skill, for example, can be assigned to a special group for that purpose. Another student who has already demonstrated proficiency in that skill is free to join a group pursuing reading or writing projects. Rarely does membership in any particular group remain the same for more than a week or two.

The value of the group lies in the peer support and learning that takes place. The success of the group generally is a clear reflection of the teacher's planning and preparation efforts.

Planning Tip: Who's Who

Any group's success depends largely on the members who compose it. Teachers who do the best with grouping usually start by giving the group a well-defined focus (often a particular skill that needs improvement) and then selecting the members.

Developing Thinking Skills Activities

In recent years, much has been written about the need to teach elementary students how to develop thinking skills. Much of this literature seems to imply that teachers have been ignoring such instruction. Nothing could be further from the truth. The problem has been that thinking skills have been taught in a very prescribed manner. In whole language classrooms, the emphasis is placed upon students being able to clarify and evaluate rather than merely restating facts.

The first step to improving student thinking skills is asking open-ended questions. For example, when reading about arctic exploration, ask questions such as "What would you take with you on an arctic expedition?" rather than, "Why did Admiral Byrd wear a parka?" Challenge students to think about possibilities rather than simply finding the correct answer in the text.

After several weeks of such instruction, teachers begin to notice subtle changes in the way students respond to one another's questions. Students think out their answers and express them more clearly. Instead of quick, flip responses, they reply with careful, honest answers. Once students begin thinking about their own thinking processes (metacognition), they are ready to move beyond the typical cognitive strategies—e.g., distinguishing fact from opinion and recognizing cause and effect—and engage in more complex thinking skills such as assessing outcomes, evaluating arguments, and probing the basis for emotional responses.

Teachers must remember that, as is the case with reading and writing skills, students differ in their critical thinking capabilities. It is important, therefore, to validate every legitimate response and to evaluate according to the individual student's proficiency rather than any artificial standard.

Student Activity

Sharing Good Ideas

Name________________________ Date________________________

The Lightbulb

How many times have you had this experience? You are watching a movie, TV show, or concert, and suddenly you come up with a great idea that would make the performance much better. In cartoons and comic strips, illustrators often show this moment by placing a glowing lightbulb above the character's head.

This activity is designed to help you capture those great ideas on paper so that you can share them with your class. Fill out the idea form below whenever you get a great idea while working with a group in the conference center or listening to someone read from the author's corner.

Date: ________________________

Reader: ________________________

Title of work being read: ________________________

____

Things you really like about the work being read: ________________________

____

____

____

____

____

____

____

Ideas that would make the work much better: ________________________

____

____

____

____

____

____

____

Student Activity

Organizing Information

Name________________________________ Date________________________

The Collecting Thinker

When we listen to someone, our minds aren't blank. We are constantly reacting to what the person is saying. We are thinking about and *evaluating* what we are hearing. Then, if asked, we offer our suggestions and opinions.

Some people form an opinion long before a speaker finishes. Such people tend to see a situation from a single point of view, and they often don't listen carefully to everything the speaker has to say. A person who has developed strong listening habits, on the other hand, collects as much information as possible *before* arriving at a conclusion or forming an opinion.

This activity will help you collect information, arrange that information in a logical order, and then develop your own opinion about what you have heard.

Listen carefully while someone speaks to you. As you are listening, make note of why you are being asked to listen.

Write down the speaker's major point(s).

What is the speaker's conclusion?

Write down all of the speaker's points with which you agree.

Write down all of the speaker's points with which you disagree.

____

____

____

____

What is your opinion of what the speaker had to say? Base your opinion on the points with which you agreed and disagreed.

____

____

____

____

Instructional Strategy

Restating Speaker's Intent

Check/Recheck

Children are not always the best listeners. Sometimes they interpret suggestions and advice as orders and directives—especially if the suggestions and advice come from a teacher or other adult or from one of the more powerful personalities in the classroom.

As a result, an off-hand suggestion can cause a young writer to change the structure, tone, and content of his or her writing. Since the teacher's role is to guide and advise rather than direct the developing writer, it's a good idea to encourage students to evaluate the advice and suggestions they receive.

During a class session or directed lesson early in the school year, have the students note some of the suggestions they have received recently. Then ask them to state why they think the suggestion was made. Finally, instruct them to accept or reject the suggestion, with a brief explanation of their reasons for taking the action they did.

Personal Planning Guide

What do you want the students to know after they complete these developing thinking skills activities?

____

____

How do you plan to present the information (chalkboard demonstration, A-V presentation, guest speakers . . .)?

____

____

What print materials do you plan to use to support learning?

____

____

Circle the features that you plan to incorporate into these activities:

demonstration application extension discussion evaluation

other: ____

Considering the nature of your students, what is the most appropriate time line for completion of these activities?

____

Complete the time line below by noting the sequence of instruction.

Step I: Introduction of material to students

____

____

____

____

____

Step II: Demonstration and modeling

____

____

____

____

____

Step III: Practice

Step IV: Collaboration/cooperation

Step V: Evaluation

Sharing as a Teaching Tool

When many of today's teachers were elementary school students, they sat in desks with straight backs arranged in straight rows. Absolute silence was the norm. Today, of course, the situation is far different. Research into how children learn has contributed greatly to the demise of such severe restrictions on student movement and expression.

Most teachers reject the notion that they are fonts of information who are obligated to spend seven hours a day lecturing students. With information expanding in geometric progression, it is impossible as well as impractical for a teacher to attempt to be the class expert in every area. No longer is the effective teacher "the sage on the stage"; now he or she is "the guide on the side." Everyone in the class is both teacher and learner.

An openness to new ideas and a nurturing of appropriate discussion techniques are common characteristics among whole language teachers. Discussion allows students to share what they know with their peers and teachers.

But discussion will not take exactly the same form from classroom to classroom. It should be molded to fit the strengths and preferences of the individuals in the class. Some children are uncomfortable speaking in front of a large group while others are natural actors and at their best when addressing the entire class (or several classes). Teachers must be sensitive to these individual differences. A child who hates speaking even in a small-group situation might be a good candidate for an individual conference in which a plan could be created to alleviate the child's discomfort.

Teacher-Directed Sharing

Despite the virtues of classroom sharing, newcomers to the whole language approach should be careful not to pull back from teacher-directed activities to such an extent that the educational program begins to lose focus. Structure is still one of the key elements in whole language classrooms. The teacher must never relinquish his or her role as coach, advisor, and guide.

During a sharing session, the teacher may be the one to pose pertinent questions. These questions may be drawn from previous discussions, students' reading, or from the sharing of one or more student's writing. Questions may focus on clarifying concepts, developing possible solutions to problems, or making connections among several types of literature (e.g., between a book currently being read and books read earlier).

In responding to such teacher-directed questions, students gain experience in exercising independent thought, making or suspending judgments, and developing appreciation for various points of view. In addition to posing the questions, the teacher participates in the sharing session as an active listener, carefully guiding the discussion away from personal attacks or put-downs and toward logical, fair conclusions.

The teacher's role in a sharing session also involves careful preparation. Before the session begins, the teacher should have designed the format that the discussion will take. Little planning is needed if a student (or students) will lead the group. If the teacher intends to have the discussion accomplish specific learning objectives, however, he or she should have a detailed outline in place before the discussion begins.

Planning Tip: Frameworks

When planning a teacher-directed discussion, first select the specific topics you want to cover and list each one on an index card. Below the topic, write three or four questions you want the students to answer. Try to develop questions that can flow easily from one to another as students give their answers. On a separate card, make a list of those students from whom you want to hear. Leave space beside each name so that you can write comments about the students as they participate.

Instructional Strategy

Questions, Predictions, and Reflections

Questioning

Asking and answering carefully posed questions are among the most difficult tasks for students to master. Teachers using the whole language approach should expect to spend several weeks at the beginning of the year helping students develop appropriate questioning skills. Then they should plan to review and reinforce these skills several times during the school year.

One effective technique is to provide students with the opportunity to pose critical questions about a variety of things that take place in the classroom. At the start of the school day, for example, have students spend a few minutes writing and then sharing *predicting* questions that focus on what they believe will happen during the day. Refer back to their predictions periodically throughout the day, and encourage students to evaluate how well they synthesized what they knew in order to make accurate predictions.

Prompt students to pose *evaluative* questions about the book they are reading. These questions should lead students to form opinions about character sentiments and motivations as well as to reflect on what happened to whom at what time. Encourage students to answer with "Yes, and also . . ." or "No, but consider . . ."; the idea is to get them to reply with more expansive responses. By using the Socratic method—answering one question by posing another—you can keep a discussion moving forward while helping students refine their questioning skills.

Consider establishing a "Tell Me More" center. Actually a community bulletin board, this center provides a place where students can pose thoughtful questions about any number of topics. A student who has specific information to share on a topic can respond to a question by posting an answer directly on the board.

Student-Directed Sharing

Many of the best sharing sessions are those directed by students with little or no input from the teacher. One example of student-directed sharing is the "Author's Chair," when a student reads a piece of writing and the class responds with a critique. While the teacher is an active participant in the session, students are free to take the discussion in almost any direction (within commonly understood boundaries).

Another example of student-directed sharing takes place when a student wishes to share with the class some of the information he or she has researched. Such sharing may be formally scheduled at the beginning or end of a writing session, or it may be an informal gathering at the conference center. No matter when the session occurs, the student leading it has the privilege of structuring it however he or she pleases.

Planning Tip: Tickets Please

By the middle of the year, several students will have compiled a great deal of information on subjects of particular interest to them. When they discover that they can't fit all of this information into their written work, they may become disenchanted with the research process and feel that much of their effort has been wasted. Simply telling them that no researcher can predict just how much information is necessary will be of small comfort to these students. Instead, consider holding an "Information Fair" at which they can share their knowledge through displays and presentations. You can invite other classes to attend the fair, and you may even wish to send "admission tickets" home to the parents of children in your class.

Developing Speaking Skills Activities

Most of us can remember a time in our young lives when we were called upon to recite a passage from memory, present an oral report to the class, or act in a school play. Some of us have fond memories of such opportunities for public speaking while many of us can still recall the anxiety we experienced at the prospect of stepping into the spotlight.

Helping students master speaking skills is one of the more difficult tasks that language instructors face. An effective public speaker must have confidence while addressing an audience. As a result, the teacher must provide emotional support and guidance as well as skill instruction. Reluctant speakers seldom find reassurance in iron-clad requirements and deadlines. Instead, they respond most positively to gentle coaching from a trusted teacher or friend.

Planning Tip: Calling on the Experts

Try to provide your students with plenty of opportunities to speak about subjects of interest to them. At the beginning of the school year, present each child with an index card with the heading "I Know All About . . ." at the top. Ask the students to write down three topics about which they are quite knowledgeable. Then collect these cards and make note of the topics and experts in your plan book. When you reach a point in the curriculum where you can make a logical connection with one of these topics, call on the expert to share his or her knowledge with the class.

Student Activity

Sharing Information

Name______________________ Date______________________

Resident Expert

It's only natural for someone who is interested in a topic to collect information about it. Some of this information will be common knowledge, but much of it will be new—and often quite interesting—to people who have not researched the topic. For example, everyone knows that all the dinosaurs are now extinct. Not everyone knows, however, that cockroaches lived during the time of the dinosaurs.

The purpose of this activity is to help you share information that you find interesting with other people. Try to develop a presentation that will inform, intrigue, and interest the people to whom you are speaking. When planning your presentation, keep one thing in mind: Don't overdo it. Simple is better than complicated.

What is the topic of your presentation?______________________

____

____

____

How long have you been studying this topic?______________________

____

____

____

Do you know any other experts in this field?______________________

____

____

____

What drawings, photographs, or other visual aids do you have that can help people understand what you are talking about?

____

____

____

____

What are the four most important things you would like people to know about this topic?

1. ______________________________

____

____

____

____

2. ______________________________

____

____

____

____

3. ______________________________

____

____

____

____

4. ______________________________

____

____

____

____

On the following page, arrange your major points in the order you would like to discuss them. Below each major point list three pieces of information that support it. Here's an example:

Cockroaches lived during the time of the dinosaurs. (Major Point)

1. Their small bodies covered with a hard shell made them adaptable. (Supporting Information)
2. They were able to eat almost anything to stay alive. (Supporting Information)
3. When the Earth became too cold for the dinosaurs, the cockroaches survived. (Supporting Information)

Major Point 1: ______________________________

____

1. ______________________________
2. ______________________________
3. ______________________________

Major Point 2: ______________________________

____

1. ______________________________
2. ______________________________
3. ______________________________

Major Point 3: ______________________________

____

1. ______________________________
2. ______________________________
3. ______________________________

Major Point 4: ______________________________

____

1. ______________________________
2. ______________________________
3. ______________________________

Instructional Strategy

Listening for Information

Guest Speakers

No one teacher has all the answers, but nearly every teacher has access to a community of experts who often are delighted to visit classrooms as guest speakers.

To make the most of a guest speaker, spend some time developing pre-visit expectations. Discuss with the students who the person is, what he or she does for a living, and what information they can expect to learn from the expert. Encourage the students to develop some questions they want the speaker to answer. Then forward these questions to your guest at least one day *before* the presentation.

Take notes during the guest speaker's presentation, and encourage students to do the same. These notes may prove very helpful in a question-and-answer session after the speaker finishes as well as in later classroom conferences and discussions.

After the speaker leaves your classroom, conduct a discussion in which students share some of the information they just learned. Challenge the students to show how what they learned might relate to their own lives. Encourage any students who are interested in further communication with the guest speaker to write letters to him or her. Advise the letter writers to include some of the highlights of the post-speech discussion—i.e., specific examples of students sharing what they had learned.

Student Activity

Stating Opinions

Name______________________________ Date______________________________

My Turn at Bat

When people speak to one another, they usually offer two kinds of information: facts and opinions. Facts are pieces of information that can be proved true. For example, evidence used against a person charged with a crime must be factual.

Opinions reflect how a person feels about a certain subject. A person can hold strong opinions, but those opinions may not be supported with facts. Effective speakers try to balance facts and opinions as they discuss their topics of interest.

This activity will help you develop your thoughts about a specific topic before making an oral presentation to the class. The topic may be something you read in the newspaper or in a book, something you saw on TV, or it may relate to an actual event in your life. No matter what your topic is, you should try to develop a clear line of thought about it. And you should try to balance facts and opinions.

What is your topic?______________________________

____

What opinions do you now hold about this topic?______________________________

____

____

Before answering the next questions, put this sheet aside for at least one day.

How have your opinions changed since you first started to think about this topic?______________________________

____

____

What facts do you have to support how you feel about the topic? Be sure to note the source for each of your facts—for example, *New York Times, National Geographic, Weekly Reader.*

____

____

____

____

Does your topic pose problems for your oral presentation? If so, what are some possible solutions to these problems?

____

____

____

Put all of your facts and opinions into an outline. Organizing your information in an outline will help you order your thoughts before you make your presentation in front of the class.

Personal Planning Guide

What do you want the students to know after they complete these developing speaking skills activities?

____

____

How do you plan to present the information (chalkboard demonstration, A-V presentation, guest speakers . . .)?

____

____

What print materials do you plan to use to support learning?

____

____

Circle the features that you plan to incorporate into these activities:

demonstration   application   extension   discussion   evaluation

other: ____

Considering the nature of your students, what is the most appropriate time line for completion of these activities?

____

Complete the time line below by noting the sequence of instruction.

Step I: Introduction of material to students

____

____

____

____

____

Step II: Demonstration and modeling

____

____

____

____

____

Step III: Practice

Step IV: Collaboration/cooperation

Step V: Evaluation

5 THE SKILLS COMPANION

Perhaps the greatest concern regarding the whole language approach—as voiced by teachers, parents, and administrators—relates to grammar and usage skills. Generally, this worry quickly disappears and is replaced by considerable support when those who voice the concern recognize that the whole language approach fosters development of these critical language skills.

The classroom in which whole language is in practice is a place in which structure is embedded in the daily routine. Students know what is expected of them, and they know the steps necessary to fulfill those expectations. Part of the daily routine consists of instruction in and reinforcement of grammar and usage skills. The teacher focuses on these skills—so necessary for academic success—without workbooks keyed to basal reading texts.

Skill instruction can be found throughout the whole language classroom. It can be found in reading and writing conferences. It can be found in the conference center, where students consult one another on ways to improve a piece of writing. It can be found around the author's chair in the critiques that students share. And it can be found at every desk as students engage in the editing process, refining their work from rough draft to final article or story. What makes the whole language approach to skill instruction so valuable is that it involves activities that are directly relevant to the student.

Students don't waste time or lose interest by reviewing material they have already mastered or attempting to acquire skills beyond their capabilities.

At the beginning of the week, the whole language teacher chooses which skills to cover—e.g., using descriptive adjectives or finding the main idea. Then, when working informally or in conferences, the teacher notes which students need assistance with those skills. The teacher then tailors instruction to meet the students' needs.

Since it is quite likely that several students share a similar need for particular skill instruction, the teacher can bring them together for topic lessons. In a whole language classroom, students enjoy the advantage of having instruction matched to their learning style and presented at a pace consistent with their ability to absorb and understand. Since the students in the skill group all need instruction in the same area, none will be intimidated by more knowledgeable peers. As a result, participation will increase, the level of commitment to learning will grow, and the degree of skill retention will rise.

Planning Tip: Skills Target

When incorporating skills instruction into your whole language program, you may find it useful to make a complete list of the skills you want (and the curriculum guide expects) students to master during the year. Once you have such a list, break it down into groups of three or four skills that seem to go together—e.g., capitalization, proper nouns, abbreviations, and main character.

At the beginning of the week (or any time frame you devise), select one of the skill groups and do an informal assessment of which students need instruction and which students show proficiency in those areas. Based on this assessment, assemble your direct teaching groups.

Repeat this procedure each time you prepare to teach new skills. Use the chart below to record which students required instruction and which ones demonstrated proficiency in particular language skills.

Date: Skill:

Students with Perceived Need	Students with Proficiency

Integration of Skills

It is impossible to force the whole language approach into the confines of a tightly structured time schedule. As teachers using the approach have discovered, the best way to implement the program is to schedule large blocks of time during which students can work in several language areas: reading, writing, listening, and speaking. Any attempt to treat these skill areas as separate disciplines will seriously compromise the success of the whole language approach.

Many teachers who are just beginning to look at whole language often ask, "But how am I going to fit in everything and be able to demonstrate to the administration that I'm covering the entire curriculum?" Teachers familiar with the whole language approach would respond by pointing out that the idea isn't to get everything to "fit in" but rather to get learning to "fit together."

In the whole language classroom, skills are not taught in isolation. Instead, they are taught and reinforced in reading and writing conferences, in small-group focus lessons, in the editing process, and in preparation of the final draft. Students examine writing styles in science, social studies, music, and art classes as well as in their language block. A basic premise of the whole language philosophy is that integrated learning is more meaningful to the student and promotes higher-level learning.

Planning Tip: One Thing Leads to Another

When planning science, social studies, math, music, and art lessons and activities, ask yourself this question: Can I integrate language skills and topics into the other content areas? By the same token, don't hesitate to make assignments in the content areas that require students to demonstrate specific language skills. You can expand this philosophy of skill integration by attempting wherever possible to relate topics from various disciplines to one another.

Methods of Implementation

Learning theories may be of help in constructing a sound educational program, but the soundness of any program surely depends most upon its practical applications. Whole language theories—the integration of skills and learning language lessons within the context of reading and writing activities—have been shown to be sound and valuable. Putting these theories into practice, however, is not always an easy task.

Here are some keys to successful implementation of a whole language program:

★ Have a clear understanding of what you want your students to know when they leave you at the end of the year. This knowledge base should include social, academic, expressive, and physical components.

★ Always be looking for those "teachable moments" when introduction or reinforcement of a particular skill is most propitious.

★ Form small groups of students who demonstrate a similar need for instruction in a specific language skill. Teach the skill to the group and then reinforce that teaching during reading/writing conferences and assignments in any of the content areas.

★ Maintain an anecdotal journal about each child. Jot down a few notes about skill development whenever you work with a child. One effective method for keeping the journal current is to make quick notes on Post-It™ paper, attach the notes to student work, and later transfer the notes to the journal.

★ Develop methods for making informal assessments of students. With some students, you may want to draw upon a particular piece of their writing or their understanding of a passage in a book. You may also find it effective to have students complete a self-evaluation of their progress.

Needs and Goals

What do we expect from our students? What do they expect of themselves? Does the community itself value learning? Do parents maintain reasonable goals and expectations for their children? Can students find information efficiently in a thesaurus, dictionary, and encyclopedia? Do the resource materials in the library support a whole language classroom?

All of these questions come into play when developing a needs/goals list for an individual student. Before such a list can be developed in the areas of grammar and usage, however, the teacher must determine what the student has already achieved. Here are several methods for doing that.

Interview

If possible, interview the teacher(s) who taught your student last year. What approach to skills instruction did they use? If they employed basal texts, which texts were they? If the whole language approach was used, what types of literature did the student favor?

Observe

Watch your student carefully to determine his or her probable learning style. When the student is reading, note his or her general tone and pacing. Does the child exude confidence or express reticence when called upon to read?

Survey

Ask both the student and his or her parents what they perceive to be the student's needs and goals. Combine your survey findings with the information you derive from interviewing and observing to shape an appropriate set of needs and goals for each student.

Collect

Collect student writing samples. Note any obvious areas in need of improvement. Do you find similar mistakes running throughout the samples?

Assess

Assimilate and correlate all the information you can gather about the student. What common traits can you detect? What major areas need improvement? In what areas might the student benefit from being challenged?

Plan

Begin to establish a plan that you can realistically implement with individual students.

Planning Tip: Dear Mom . . .

When you first implement the whole language approach in your classroom, be sure to inform the parents about the strengths of the program. Since most parents are quite concerned about skill instruction, let them know which skills are most appropriate for their children to be learning. Give parents as many examples as possible—e.g., "If your child is having difficulty recognizing proper nouns, we will address that need in focus lessons, individual and group conferences, and editing sessions." Point out that the whole language approach to skills instruction increases student interaction with language in a manner that is more personalized and more engaging than could be experienced with standard workbook pages.

Parts of Speech

In many traditional language arts classes, more hours are spent on the parts of speech than any other topic. Each year, students are reintroduced to "The Noun: Keystone of the Sentence," "Our Friend the Verb," and "What Would We Do Without Adverbs?" They listen to lectures, see examples, and complete hundreds of exercises in their attempt to master the parts of speech. Although teachers have been using this approach for decades, there is little evidence to show that it leads to marked improvement in the way students learn English.

The whole language approach infuses the parts of speech into the daily classroom routine. This infusion takes three forms: direct instruction during focus lessons, student-teacher conferencing activities, and student-directed activities. The two sections that follow cover conferencing activities and student-directed activities.

Planning Tip: That's Progress

Students should always have a clear idea of the skill(s) on which they are supposed to be focusing. Therefore, as soon as you select the skill you are going to focus on with a student, be sure to inform him or her during a conference. Then check periodically to make certain that the student is indeed focusing on that skill.

When both you and the student believe that enough time has been spent focusing on the particular skill, invite the student to assess his or her improvement. This self-evaluation may take any of several forms: a time line, a comparison between a previous writing sample and one completed toward the end of the focus period, or a simple description of what the student has learned. Encourage the student to share this self-evaluation with his or her parents.

In the whole language classroom, students should be allowed to become active participants in the evaluation process.

Parts of Speech Conferencing Activities

In the whole language classroom, students are active participants in the learning process. They are constantly aware of both their success and those areas in which they need to improve.

Accordingly, the teacher who decides that a student needs instruction regarding a part of speech shares that decision with the student during an individual conference. Teacher and student develop a time line for attaining a specific improvement goal and agree upon ways to focus on that part of speech in reading and writing. The teacher then introduces the student to other members of the focus group.

Teacher and student hold a second individual conference at a later date to assess the improvement made and to look at ways that the part of speech appears both in the student's writing and in the books he or she is reading. The books serve as a model for the student's writing, and the teacher encourages the student to develop strategies for using the part of speech as he or she sees it used in the published works. The teacher or, preferably, the student should make a list of those strategies.

From time to time in the whole language classroom, the teacher holds group conferences during which all of the students focus on a specific part of speech. The first session of such conferences should include direct teaching of the strategies needed to identify the part of speech, its purpose, and the form it takes in writing. Naturally, the teacher must adjust the instruction level to the abilities of the students in the group. Some groups may require just one direct teaching session while other groups may need several. In determining the amount of time each group needs for such instruction, the teacher makes constant assessments of student abilities through reading, writing, and discussion.

The following activities are designed for use in individual and group conferences.

Instructional Strategy

Nouns

Who and What

If a student is having difficulty using nouns or experiencing confusion as to placement and form, those problems will likely be most evident in his or her writing. The teacher can help the student through direct teaching in a small group or during reading and writing conferences. The advantage of the conference session is that it generally takes on a more personal tone.

When conferencing with a student and focusing on the noun, challenge the student to become a "scene painter." Have the student list the people, places, and objects that appear in a scene from the book he or she is reading or from the article or story he or she is writing. Start by having the student visualize the scene in his or her mind. Then give the student a large sheet of paper. Instead of sketching the scene itself, however, the student must place the nouns on the sheet where each person or object would appear.

When the student finishes, he or she then gives the sheet to two other students in the class. They each paint what they believe the scene looks like based strictly on the first student's "noun picture." These two students must not talk to each other or consult with anyone else in the class. When the students complete their scenes, compare the two pictures. The differences illustrate the idea that common nouns (such as "car") can be portrayed many ways while proper nouns (such as "Jaguar") take on a more specific appearance.

You can also use this same activity for adjectives.

Instructional Strategy

Verbs

Lights, Camera, Action

Young readers seem to enjoy books with plenty of action. Although they have little difficulty describing what happens in a story (often in great detail), some of these children cannot identify the part of speech that puts the characters in motion.

One strategy likely to help students who have trouble identifying verbs involves freezing the action in the story they are reading. Ask a student to read a portion of the story. After a few sentences suddenly say, "Stop!" After the student recovers from the initial shock of the command, explain that nothing will happen until the two of you start the action again.

Hand the student a sheet of paper, and ask him or her to list as many words as possible that might get the action of the story going again. For example, does Karen *walk* out of the room or does she *storm* from the room? After the student writes down several words, discuss how the verb not only puts the characters and events in motion, but how it also flavors the action that takes place.

Challenge the student to read something else (it may be something you have written, a passage from a book, or a sample of the student's own writing). After he or she has read the piece once, have him or her reread it, this time replacing the verbs with other ones that might add a bit more flavor. As the student is substituting verbs, informally check for verb recognition, noun-verb agreement, and the student's willingness to take risks.

Instructional Strategy

Adjectives

Not Just Any Car

When students read, they learn to recognize and appreciate adjectives. When they write, however, students tend to overuse adjectives or they keep repeating the same adjectives over and over again. To the overusing student, every car becomes a shiny red antique smoky rattling convertible. To the student with only a small pool of adjectives to draw upon, the fast dog chases the fast car past the fast runners down the fast street.

You can help such students develop a vocabulary of appropriate adjectives and use them in a balanced manner. One way you can help is by initiating an adjective bank. Place several common nouns at the top of a large sheet of paper. Challenge students to list fresh, descriptive adjectives below each noun. Encourage them to avoid overused adjectives like "big," "small," "fast," and "slow."

When they finish, tell the students that they may borrow words from this list whenever they need to enrich their writing. But every time they use one of these words they must repay the bank by adding two more and signing their name to the list.

This activity not only helps students build a vocabulary of adjectives, but it also encourages them to use descriptive words in their writing that they may never have considered without the use of the adjective bank.

Instructional Strategy

Adverbs

Like a Slowly Rolling Stone

Helping students recognize and use adverbs is often a formidable task. To the young writer, adverbs can be elusive.

One effective way to improve student recognition and use of adverbs is to show how adverbs describe *time, place, manner*, and *degree*. Adverbs of *time* provide information about when, how often, and how long. Adverbs of *place* provide information about where. Adverbs of *manner* explain how something is done while adverbs of *degree* explain how much or how little.

When conferencing about adverbs, hand the student a chart like the one below. Have the student skim a passage from a book or a short piece of his or her own writing, looking for any words that could be placed in the boxes. Assist the student in placing each adverb in the correct quadrant, and then review the four major groups.

ADVERBS OF TIME (provide information about when, how often, and how long)	**ADVERBS OF MANNER** (explain how something is done)
ADVERBS OF PLACE (provide information about where)	**ADVERBS OF DEGREE** (explain how much or how little)

Personal Planning Guide

What do you want the students to know after they complete these parts of speech conferencing activities?

____

____

How do you plan to present the information (chalkboard demonstration, A-V presentation, guest speakers . . .)?

____

____

What print materials do you plan to use to support learning?

____

____

Circle the features that you plan to incorporate into these activities:

demonstration application extension discussion evaluation

other: ______________________________

Considering the nature of your students, what is the most appropriate time line for completion of these activities?

____

Complete the time line below by noting the sequence of instruction.

Step I: Introduction of material to students

____

____

____

____

____

Step II: Demonstration and modeling

____

____

____

____

____

Step III: Practice

Step IV: Collaboration/cooperation

Step V: Evaluation

Parts of Speech Student-Directed Activities

The activities in this section, designed to assist students in focusing on specific parts of speech within the context of their own reading and writing, can be employed in a variety of ways. They may be used in the conference center, supporting the preparation of author's chair material, as part of the final editing process, or as a free-time activity.

Students should decide how best to utilize these activities, and they should be the ones to confirm that the activities have been completed. Teachers should note student progress during individual conferences and perhaps use such progress as an informal measure of student achievement toward the goals spelled out by the student-teacher team.

All of these activities can be completed by individuals or by groups. Sometimes peer teaching and support can be valuable resources in student mastery of parts of speech. If the particular mix of students in a group situation might compromise the stated goals, however, then the individual approach would be advisable.

Student Activity

Nouns

Name________________________________ Date________________________________

What A Character

It happens to everyone from time to time. You start reading a book and find yourself having trouble keeping the characters straight. Is it Frank who lives in Phoenix, or is that Jason? Does Heather drive a Ford Bronco II, or is it a Jeep CJ? Is Francis a man or a woman?

Part of the problem is that all of these people, places, and things are *nouns*. Nouns, as you probably already know, are the part of speech that names people, places, things, and sometimes ideas. There are basically two kinds of nouns. *Proper nouns* refer to specific people, places, and things—Ford Bronco, for example. *Common nouns*—such as truck—are less specific.

In this activity, you role-play a news reporter. Your job is to collect as much information as possible about the characters in the book or story you are presently reading or writing. Make up a chart like the one below for each of the characters, and then fill in the information.

Character's name: ________________________________

____

Place where character lives: ________________________________

____

Other people with whom character lives: ________________________________

____

Other characters with whom this character associates: ________________________________

____

____

____

This character's most prized possessions: ________________________________

____

____

____

Now, look carefully at the information in your chart. Are all the proper nouns listed accurately? Are all the names spelled correctly? Place a check mark beside each proper noun in your chart.

Student Activity

Verbs

Name____________________________ Date_______________________

And Then . . .

How many times have you heard someone describe something that happened, and you know deep down that most of the story is an exaggeration? People often tell jokes about "the fish that got away" (it's always bigger than any fish they caught) or about their "exotic vacation" (spent in a cabin at a nearby lake).

These "tall tales" often involve more than exaggerated stories. They usually involve exaggerated language as well. For example, the car did not *drive* down the street—it *zoomed*. The stranger did not *stand* in the alley—he *lurked*. The italicized words are *verbs*. Verbs express the action in a story.

You can create your own tall tale in this activity. Start by selecting a passage from the book you are reading (one or two paragraphs may be all that you need). Find all the verbs in the passage, and replace as many as possible with ones that make the story a wild exaggeration. When you finish, read both the original passage and your version to several of your friends. Can they tell which piece of writing is from the book and which is your tall tale?

You may want to illustrate your exaggerated story with original drawings.

Student Activity

Adjectives

Name________________________ Date________________________

The Great, Big, Wild . . .

If you're looking for a way to create a terrible, boring, uninteresting piece of writing, here's one guaranteed way to do it: leave out every adjective. Suddenly, "The ancient woman carrying the tattered and scuffed bag" becomes "The woman carrying the bag."

Adjectives are an important part of speech. They bring color, size, age, and other descriptive words to your writing, allowing you to write with power and expression.

One of the best ways to become skillful at recognizing and using appropriate adjectives is to examine how your favorite author uses them. Select a passage from a book you like very much and look for all the adjectives. How do these words affect the story? How much less interesting would the story be if the author had not used them?

The purpose of this activity is to help you evaluate the use of adjectives in the book you are reading. On the line at the top of each box, write the name of a character, a character's pet, or some object that the character uses in the book. Then try to fill up the box with as many adjectives as you can find in the book that describe that person, pet, or object.

____

____

Personal Planning Guide

What do you want the students to know after they complete these parts of speech student-directed activities?

____

____

How do you plan to present the information (chalkboard demonstration, A-V presentation, guest speakers . . .)?

____

____

What print materials do you plan to use to support learning?

____

____

Circle the features that you plan to incorporate into these activities:

demonstration application extension discussion evaluation

other: ____

Considering the nature of your students, what is the most appropriate time line for completion of these activities?

____

Complete the time line below by noting the sequence of instruction.

Step I: Introduction of material to students

____

____

____

____

____

Step II: Demonstration and modeling

____

____

____

____

____

Step III: Practice

Step IV: Collaboration/cooperation

Step V: Evaluation

Usage Conferencing Activities

"That don't sound right to me."

This expression of confusion is an obvious clue that a student is not familiar with correct English usage. Helping a student overcome such confusion often isn't easy, requiring a good deal of practice, guidance, and self-evaluation. The teacher's role is to clarify and promote as much as possible correct usage in student oral and written language skills.

The following activities are designed to assist the teacher perform that role. As was the case with parts of speech, the usage activities are divided into two sections: conferencing activities and student-directed activities. The conferencing activities may be employed during individual or group conferences.

Planning Tip: Speak With the Pros

Nearly every community has language professionals, people who make their daily living through reading, writing, and/or speaking. You can call upon these people to visit your classroom. Local newspapers, radio, and television stations are often pleased to send their celebrities to schools to discuss how the media work. In addition to the well-known personalities, you can invite their editors and program directors.

Language professionals can be put to work in the conferencing center, helping students polish a piece of writing or prepare for a presentation. If you can establish mentor relationships between these pros and some of your reticent readers, writers, and speakers, you may well see dramatic changes in student development and achievement.

Instructional Strategy

Vocabulary Selection

Teaching Old Words New Tricks

Language lives. As a society changes, so does its language. Just 25 years ago, the word "chip" had little use outside of its association with Las Vegas and potatoes. Today, that word is more commonly used in conjunction with computers and electronic games. Helping young writers understand the changing nature of language is not easy. Even harder is getting them to adjust and conform to the demands such a changing language imposes.

One useful strategy involves challenging students to find new words and new meanings for old words. Ask your students to collect words that they think did not exist or had a different meaning a generation ago. They can list the words on large wall charts or deposit word cards in coffee cans. Encourage the students to classify the words according to parts of speech. Then, every quarter or so, have the students alphabetize the words and add them to an ongoing classroom dictionary of new terminology.

Instructional Strategy

Dialect and Characterization

In Character

A Texas cowboy would probably not greet a stranger with "Top of the morning to ye" any more than a leprechaun would proclaim, "Move along little doggies."

When students write, they often use their own voice in dialogue, little realizing that doing so may be inappropriate to the way their characters should speak. In this activity, which works most effectively in a large-conference or whole-class format, you can help students experiment with dialect and characterization.

Introduce the class to dialect by recording yourself saying several lines in a number of regional dialects. Then sketch pictures of what each speaker might look like. You might want to ask for the assistance of an artist at this stage of the activity. Play the tape and show the pictures to the class. Have students try to match the voice with the person.

Once students seem to understand the idea of character voice, have them begin adding dialects to the dialogue in their stories.

Personal Planning Guide

What do you want the students to know after they complete these usage conferencing activities?

____

____

How do you plan to present the information (chalkboard demonstration, A-V presentation, guest speakers . . .)?

____

____

What print materials do you plan to use to support learning?

____

____

Circle the features that you plan to incorporate into these activities:

demonstration   application   extension   discussion   evaluation

other: ____

Considering the nature of your students, what is the most appropriate time line for completion of these activities?

____

Complete the time line below by noting the sequence of instruction.

Step I: Introduction of material to students

____

____

____

____

____

Step II: Demonstration and modeling

____

____

____

____

____

Step III: Practice

____

____

____

____

____

Step IV: Collaboration/cooperation

____

____

____

____

____

Step V: Evaluation

____

____

____

____

____

Usage Student-Directed Activities

Most of the following activities deal with the editing process. It is at the editing stage that students should put into practice the techniques they have learned through conferencing and practice.

Perhaps the most effective way a teacher can help a student who has displayed some difficulty with usage is to pair him or her with a student editor who has shown proficiency in that area. The idea is not to have the editor serve as a grader but rather as a peer who can pose valuable questions. By answering the editor's questions, the student writer is likely to improve his or her article or story.

Encourage students to compare and contrast their writing with models of good writing that they selected in earlier reading and conferencing sessions. Help them develop a mental checklist for use in evaluating their own writing. Most importantly, have them go back to articles and stories they wrote some time ago and look for usage mistakes they no longer make. This technique is sure to motivate them to continue to improve.

Planning Tip: Editorially Speaking

Although student editors can be a real plus in your classroom, these young people must be sensitive to the strengths and weaknesses of their peers. Some teachers have would-be editors fill out an application and go through a training session before being paired with student writers. The application should be easy to fill out, and the training session should emphasize the need for sensitivity in working with others. Be sure to remind editors to refer questions they cannot answer to another editor or to you; they must not try to act "all-knowing." In addition, help them to prepare sample time lines that they can use when working with their peers.

Student Activity

Measuring Success

Name________________________ Date________________________

I've Learned

After you have been reading and writing like an author for several months, you suddenly realize that you are able to read and write so much better than before. You begin to see yourself as a writer—someone who is able to work with words, someone able to use words to make other people laugh and cry, someone able to communicate well with others through writing.

In addition, you have gained many practical writing skills. You probably didn't really understand what a paragraph was until you began to write. You likely also found it difficult at first to create engaging dialogue. But now you can do all of those things and more!

After working hard to develop your language skills, it's nice to sit back, take a look at everything you've learned, and appreciate the progress you've made. Here's a chart to keep in your folder. Each time you think of something new that you've learned or a skill that you've mastered this year, fill in one of the lines on the chart.

DATE **I'VE LEARNED . . .**

DATE	I'VE LEARNED . . .

Personal Planning Guide

What do you want the students to know after they complete these usage student-directed activities?

____

____

How do you plan to present the information (chalkboard demonstration, A-V presentation, guest speakers . . .)?

____

____

What print materials do you plan to use to support learning?

____

____

Circle the features that you plan to incorporate into these activities:

demonstration application extension discussion evaluation

other: ______________________________

Considering the nature of your students, what is the most appropriate time line for completion of these activities?

____

Complete the time line below by noting the sequence of instruction.

Step I: Introduction of material to students

____

____

____

____

____

Step II: Demonstration and modeling

____

____

____

____

____

Step III: Practice

Step IV: Collaboration/cooperation

Step V: Evaluation